Computer-Aided Design of

Feedback Control Systems for Time Response

Computer-Aided Design of

Feedback Control Systems for Time Response

Clarence J Maday

Kogan Page

First published 1987 by The
Instrument Society of America,
67 Alexander Drive, P.O. Box 12277,
Research Triangle Park, NC 27709, USA.

Published in Great Britain by Kogan Page,
120 Pentonville Road, London N1 9JN

British Library Cataloguing in Publication Data

Maday, Clarence J.
Computer-aided design of feedback control systems for time response.
1. Automatic control – Data processing
I. Title
629.8'312 TJ213

ISBN 1 85091 421 4

Printed in Great Britain by
Billing & Sons Ltd., Worcester

Contents

List of Figures

List of Tables

Preface

The material in this "how-to" book builds upon a first course in classical control. Appropriate concepts from modern control are used. Explanations are based upon the author's insight and intuition gained through his experience with dynamical systems. In particular, the author tries to show how error compensation networks are developed. In most classical control courses, this topic is usually scheduled near the end, which often means it is not covered. This problem is addressed directly here because the subject of error compensation unifies the presentations in the book.

Frequency domain analysis herein is limited to pole assignment. Time domain analysis through numerical simulation dominates the presentation. In linear systems, the solution to the eigenvalue problem gives only a glimpse of the system response. In large systems, we can determine stability and a bit about oscillatory behavior. The usual worth of a picture — 1000 words — certainly holds true here. In nonlinear systems, numerical simulation may be the only available tool. The first system in Chapter 3 is nonlinear and emphasizes this. This is also true for linear digital systems where the sampling frequency is only two to five times greater than the highest system frequency. Such cases are described in Chapter 5. Frequency domain analysis for digital systems generally provides only little insight into the transient behavior of the system. The situation becomes virtually intractable when one tries this approach to evaluate the robustness of the system.

The material in Chapter 6 on the error observer has not been published or described before. Beyond its novelty, the error observer establishes a neat connection between classical control and modern control.

Chapter 3 is similar to "Synthesis and Design of Feedback Control Systems for Time Response I" presented at the American Control Conference in 1984 and printed in the Proceedings, pp 1332–1337. These proceedings are copyrighted by American Automatic Control Council and are distributed through IEEE Service Center, 445 Hoes Lane, Piscataway, NJ 08854.

Clarence J. Maday
Raleigh, NC

CHAPTER 1

Introduction

Historical accounts trace the use of self-regulating or feedback control systems at least as early as the third century B.C. Given the ingenuity of man, we can conjecture that earlier uses may have gone unnoticed because the inventor neglected to document such efforts. The recorded history of automatic control, however, is fascinating, and the interested reader can refer to one of the available monographs for a detailed account. We note only that today we tend to think primarily in terms of classical control and modern control. The topic usually called classical control was developed in a period that extended from the late 1930's to the early 1950's. Pontryagin's *Maximum Principle* renewed interest in state-space methods, and the general availability of high speed digital computers contributed to the development of what is called modern control theory. It has been noted in the literature that the term *modern* may be somewhat inappropriate because state-space or state variable techniques date back to the work of Poincare, Gibbs and Lyapunov around the beginning of this century. Nonetheless, the application of these methods to the analysis of control systems has proven to be valuable, and there is every indication that new developments will continue. It is generally agreed that *modern control* includes the state variable methods just indicated, as well as optimal control and digital control.

Optimal control has given us the linear quadratric regulator (LQR), and its counterpart, the linear quadratic servomechanism, both of which use complete state feedback. Difficulties are encountered when one or more of the states are inaccessible, i.e., they cannot be measured in the physical setting. The concept of the observer, which synthesized the missing states, was put forward to overcome this problem.

Digital control or sampled-data systems experienced a surge of interest during the 1950's when much of the theory was developed. Interest in these systems is renewed at the time of this writing because of the availability of very high speed microprocessors. Moreover, digital systems offer the potential of handling numbers with precision and with accuracy.

The need for feedback in electromechanical systems, chemical process control, aerospace guidance systems, internal combustion engines, and precision machining is accepted today. Feedback control systems are widely used and may be taken for granted. In view of this, it seems appropriate to reexamine the purpose of a feedback controller and to question our expectations of such a system. In short, what do we expect of an automatic feedback control system? For single-input, single-output (SISO) systems we expect the feedback system to be better than the sum of its constituent components. That is, the closed-loop performance of a controller should be superior to its open-loop performance. We expect also that the system will track faithfully any reasonable input command. Moreover, it should continue to perform well with approximate models of physical systems, and it should be insensitive to external disturbances; that is, it should be robust.

Let us consider a seemingly simple example that highlights these requirements and also serves to introduce more sophisticated concepts. The temperature control of a house is the example. The control element is an on-off heating/cooling device controlled by a thermostat, which is also an on-off device, often with a thermal anticipatory provision. The model of the house is generally taken to be a first-order one whose elements are known only approximately. Outdoor temperature variations and opened doors and windows are external and usually unpredictable disturbances. Yet the system works well even if it does not lend itself to linear analysis. We keep in mind the on-off character of the controller as a desirable feature for a controller because this means that we can use the controller at full capacity when it is needed. In economic terms, this means that we can use all of the controller that we paid for. This contrasts with our experience with ordinary linear systems where a controller cannot be driven to saturation. Thus, a larger capacity actuator is required in ordinary linear systems than would be the case in the on-off system. *Saturation operation, a nonlinear phenomenon, is the first desirable feature of a control system*. This problem is considered in Chapter 3 where it is shown that this nonlinear character can be represented by a piecewise linear system with gain scheduling. Gain scheduling is usually taken to mean feedback gains and compensation gains can be changed dynamically to achieve some desired time response. This is very much like the general linear quadratic regulator whose feedback gains are time dependent.

While the temperature control of a house is a relatively simple matter to implement, this example leads us to consider a similar but more sophisticated control problem. We consider the control of temperature and relative humidity in an environmental test chamber such as is used to evaluate refrigerator performance. The temperature and the relative humidity are coupled thermodynamically so that a change in one affects the other. If, for example, it were required to increase the temperature in the chamber, the relative humidity would decrease as the temperature increased unless water were added to the system. Even then, we can expect some transient fluctuations in the relative humidity and in the temperature before steady state conditions are attained. There are processes where these transients are undesirable and must be minimized. *Noninteracting response in coupled systems is a second desirable feature for a controller*. This is described in Chapter 4.

Saturation operation and noninteraction are features that characterize a system's time response, rather than its frequency response. This suggests that it is appropriate to determine how to achieve some desired time response. It has been found that time response is governed at least by the closed-loop poles (roots of the characteristic equation) and by the amount of integral action, i.e., the I in the traditional PID controller. There are other features that govern time response, but pole location and integral error compensation are sufficient for our purposes. Pole location establishes the nature of the response to a step input or to a ramp input. Integral error compensation provides robustness and noninteraction in otherwise coupled systems.

Pole assignment in multivariable systems can be achieved with complete state feedback. The addition of integral compensation with unspecified gains preserves pole assignment capability while providing us with the robustness just mentioned. *The third desirable feature of a control system is pole assignment with integral error compensation and state feedback*. The cost of all of this may be very high amplifier gains and noisy signals. Analog circuits become less and less attractive at this point because the signal-to-noise ratio may be decreasing dramatically. With this in mind, we consider both analog systems and digital systems in design analyses so that the designer may select the system appropriate to his needs.

This brings us to the very important problem of pole assignment in digital systems or, more precisely, in systems with a time-continous plant governed by a Direct Digital Controller (DDC), i.e., a digital computer. Pole assignment techniques are straightforward in analog-only systems and in discrete-only systems. The problem here is to represent the mixed system in a single mode, either analog or discrete. Design techniques for mixed systems are treated in detail in Chapter 5.

Integral error compensation with state feedback, actuators driven to saturation, and noninteracting response in coupled systems characterize high performance feedback control systems. In general, these features can be obtained through pole assignment. For very large systems, however, available algorithms may not be able to handle ill-conditioned systems that are often encountered. An alternative is the expansion of the system determinant in algebraic form. The required algebraic manipulations are burdensome and tedious. Fortunately, symbolic languages can be used to carry out algebraic as well as arithmetic operations. Generally available languages include REDUCE™ (Rand Corp.®) and MACSYMA™ (Symbolics, Inc.®).

The ready availability of very high speed digital computers that perform algebraic and arithmetic operations provides the opportunity to obtain the features that characterize high performance control systems, particularly those of high order. This new approach to control system analysis, synthesis, and design is based upon a combination of appropriate features from classical control theory and from modern control theory. These features are used as a foundation on which to build and to extend the results of previous investigations. There is little or no temptation to call this work by some new name, such as ultramodern or post-modern control theory. Rather, we note that this work is a partial unification of concepts from the classical approach and the modern approach. Indeed, if there is any suggestion for a new name, the concepts of unification and coherence should be considered seriously.

The emphasis in this monograph will be on the time response of feedback control systems subjected to arbitrary disturbances and a range of inputs. It is the author's intent to take full advantage of high speed digital computers for algebraic manipulation and for numerical simulation.

Previous Work

Here, we consider those established features of classical control and of modern control that will be useful in this work. System block diagrams have always had great value in synthesis and design; they continue to be useful for present purposes. System transfer functions in expanded form with algebraic coefficients are closely identified with classical control but have found application in modern control as well. Proportional control in classical terms is similar to state feedback in the modern sense. Integral compensation is sometimes referred to today as dynamic feedback control. The PI control strategy is labelled Integral Error plus State Feedback (IESF) in this monograph.

The vector/matrix methods of modern control are very useful for our purposes. From modern control we note also that some results from optimal control will be used. Time response characterizes these results; we follow these lines. Bang-bang control derives from optimal control but finite horizons must be specified. This presents a problem for indefinite horizon systems such as those subjected to arbitrary disturbances, but we keep bang-bang performance in mind as a desirable goal.

One of the more attractive aspects of the vector/matrix form of the system differential equations is that it emphasizes that physical systems consist of first-order systems and second-order systems. State variables usually suggest themselves, i.e., temperature, pressure, velocity, position. Immediately, this establishes the nature of the block diagram where each constituent is shown individually and the state variables are identified.

Sampled data or digital systems have been analyzed since the 1950's. Interest in these systems has increased with the availability of the new and very fast microprocessors. These systems are characterized by time delays required to make digital calculations or to take measurements in digital form. A difficulty associated with these systems is the usual presence of a time-continuous plant. This means that part of the system is described by differential equations (for the time-continuous plant) and the remainder of the system is described by difference equations (for the digital controller and the digital sensors). In Chapter 5 we present a method to reconcile the differential/difference character of the system.

We mentioned earlier that some state variables may be inaccessible, or impossible to sense or measure, and they prevent the implementation of a control algorithm that depends upon complete state feedback. The usual approach around this problem is to use an observer to reconstruct or synthesize the missing state variables. The traditional observer uses available state feedback to construct an estimate of the complete state. Reduced-order observers can be used to reconstruct only those variables not accessible for measurement. Although the observer poles are independent of the closed-loop system poles, the observer poles are usually picked a little to the left of the system poles.

In Chapter 6 we describe a control algorithm with a built-in "reduced-order observer" that implicitly reconstructs state variables and permits pole assignment. Since the closed-loop system poles include the observer dynamics in this case, all poles are shared by the system and the observer.

The overall objective of this monograph is the presentation of an efficient and general procedure for the analysis, synthesis, and design of multi-variable analog and digital feedback control systems that provide specified time response, robustness, and tracking ability.

Four intermediate steps that will lead to this objective include:

1. Development of a piecewise linear, robust control algorithm that achieves constraints on the state and control variables through error compensation. This allows for nonlinear response of the actuators (driven to saturation for much of a cycle) within a linear feedback framework.
2. Development of control algorithms for multivariable systems that reduce or eliminate interaction due to coupling without the use of decoupling filters.
3. Development of an algorithm to achieve arbitrary pole assignment in a mixed control system; that is, a system with a time-continuous plant to be controlled with a discontinuous digital controller.
4. Development of a control algorithm with a built-in reduced-order implicit "observer" to reconstruct or synthesize some information about those state variables that cannot be measured.

CHAPTER II

The Integral Error with State Feedback Method

In the previous chapter we noted certain concepts of classical control and of modern control that will be useful in this monograph. It is our intent to unify these concepts and extend them in a coherent manner to achieve the goals mentioned earlier.

We continue to use the transfer function/block diagram representation of system components because state variables are readily identifiable in these diagrams. Moreover, this emphasizes that large scale physical systems are an assembly of first-order systems and of second-order systems.

From the linear quadratic regulator (LQR) of modern control, we use the idea of complete state feedback. Where one or more state variables may be inaccessible for measurement, an "observer" can be employed. From classical control, integral error compensation is used as an essential part of this unified approach. We build upon the established results from classical control and from modern control to achieve new results in noninteraction, constraints on state variables, reduced-order implicit "observers," and pole assignment in combined discrete-continuous systems.

We will describe the possible forms that IESF may take with illustrative examples that include a first-order and a second-order analog plant and a digital system with a first-order plant. In Chapter 5 we treat a second-order continuous plant controlled by a second-order direct digital controller.

The first example is a single-input, single-output first-order analog plant with state feedback only (Figure 2-1). The closed-loop response transform is

$$c = \frac{Kr + T_L}{s + a + K} \tag{2-1}$$

which shows that K can be selected for closed-loop pole assignment. In this and in subsequent diagrams, T_L is a disturbance. This form of state feedback is recognized as the classical proportional controller. The Final Value Theorem shows that a finite steady-state error will result for a step input at r at the disturbance T_L. This controller can be improved by adding a stage of integral compensation as shown in Figure 2-2. The transform of the response is

$$c = \frac{K_1 r + sT_L}{s^2 + (K_2 + a)s + K_1} \tag{2-2}$$

where K_1 and K_2 can be selected for closed-loop pole assignment. We note that a step input at r or T_L results in zero steady-state error, a significant improvement over the previous example.

State feedback only is shown for the second-order (two state variables) single-input, single-output (SISO) system in Figure 2-3. The closed-loop response transform is

$$c = \frac{K_1 r + T_L}{s^2 + s\,(K_2 + a + b) + (K_1 + K_2 b \times ab)} \tag{2-3}$$

which shows that K_1 and K_2 can be selected to achieve pole assignment. Zero steady-state error for a step input at r is guaranteed, however, only for $K_2 = -a$. We see also that a step input for the disturbance T_L produces a finite steady-state error in the response. These problems with steady-state response can be treated with a single stage of integral error compensation as shown in Figure 2-4. Now the closed-loop response transform is

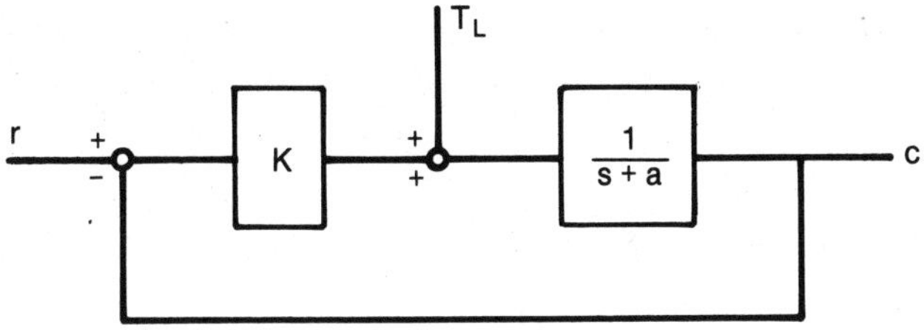

FIGURE 2-1
Single-input/single-output first-order analog plant with state feedback only.

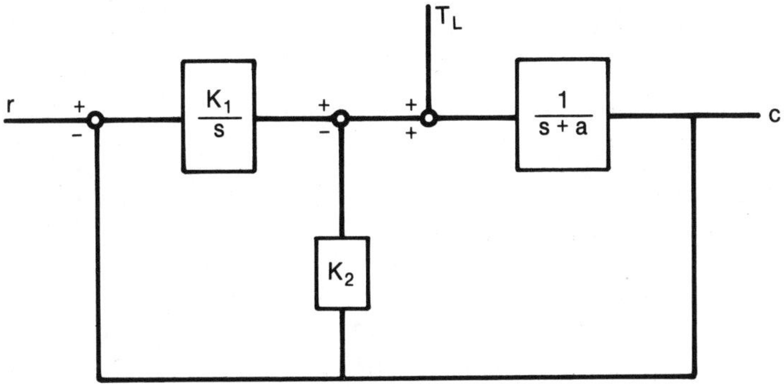

FIGURE 2-2
Single-input/single-output first-order analog plant with state feedback and one stage of integral compensation.

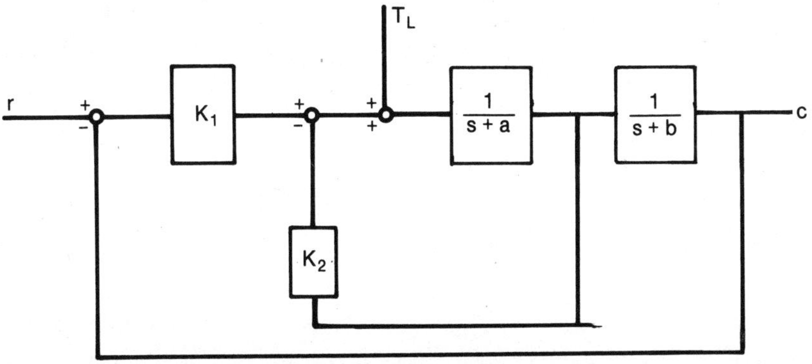

FIGURE 2-3
Single-input/single-output second-order analog plant with state feedback only.

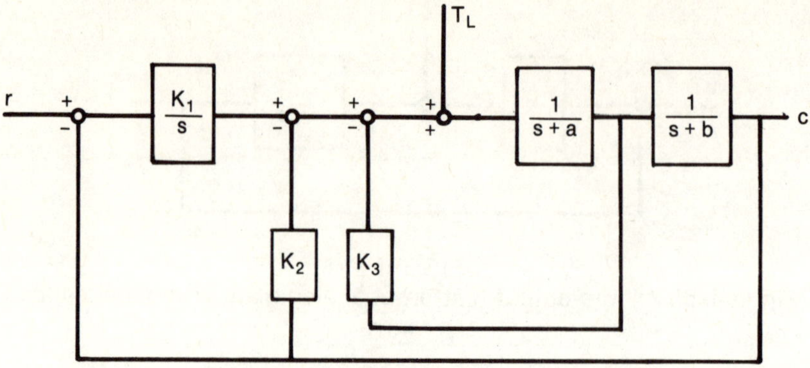

FIGURE 2-4
Single-input/single-output second-order analog plant with state feedback and one stage of integral compensation.

$$c = \frac{K_1 r + sT_L}{s^3 + s^2(K_3 + a + b) + s(K_2 + K_3 b + ab) + K_1} \qquad (2\text{-}4)$$

The gains K_1, K_2, and K_3 can be selected for pole assignment. Use of the final value theorem shows that zero steady-state error can be obtained for a step input either at r or T_L. These are attractive features. Even more effective disturbance rejection can be obtained with the addition of another stage of integral compensation as shown in Figure 2-5. Here we have what is usually called a cascaded controller, i.e., a controller within a controller. The dashed lines enclose what could be regarded as a derivative or velocity controller. The output of the outer system at point 2 is the input to the inner system where the "error" may be taken as the output of summing point 3. The transform of system response is

$$c = \frac{K_1 K_3 r + s^2 T_L}{s^4 + s^3(K_4 + a + b) + s^2(K_3 + K_4 b + ab) + s(K_2 K_3 + bK_3) + K_1 K_3} \qquad (2\text{-}5)$$

We use the cascaded controller in multi-input, multi-output (MIMO) systems with strong coupling. Multiple integrators reduce or virtually eliminate interaction effects among the different loops without using traditional decoupling filters. This is treated in detail

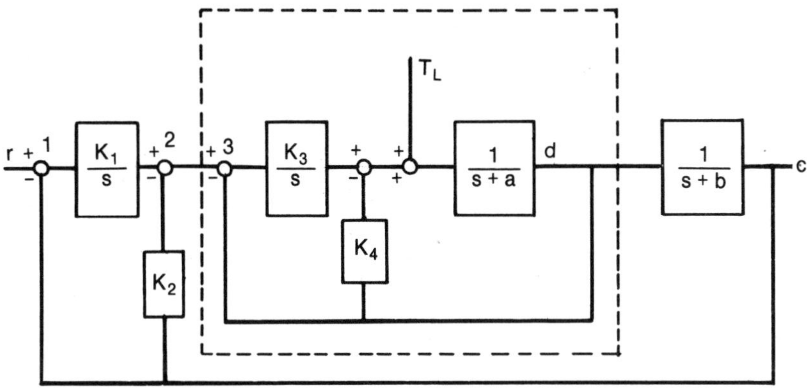

FIGURE 2-5
Single-input/single-output second-order analog plant with state feedback and two stages of integral compensation.

in Chapter 4. This form of error compensation has been called Pseudo-Derivative Feedback by Phelan (1). It can be seen from the illustrative examples that the algorithm is part of the larger class of algorithms that use state feedback with one or more stages of integral compensation.

Cascaded controllers also permit contraints on the state variables as well as the control variables through error compensation alone. In Figure 2-5 the state variable *d* can be constrained by limiting the output of point 2. This is equivalent to a step input at summing point 3 to move state variable *d* through the magnitude of the step. This has been demonstrated successfully for a third-order plant with cascaded limiting and is described in Chapter 3.

In all of the foregoing we have focused on analog systems describable in the Laplace or *s*-domain. Even though our intent is to establish design procedures for digital systems, we recognize that the general familiarity of most practitioners with analog systems justifies this emphasis here. The extension to digital systems is covered in Chapter 5.

This approach to the synthesis and design of automatic control systems is attractive because it allows the designer to obtain a desired time response for some input. For larger systems, however, there is the problem of carrying out the operations needed to obtain the algebraic expansion of the system determinant. Symbolic and algebraic manipulation (SAM) languages can help with this problem.

For comparison we show a traditional PID controller in Figure 2-6. We note that compensation is applied to the error signal only and not to state feedback information. A practical difficulty is the need to differentiate the error signal. This is usually a noisy process that can introduce spurious error signals that degrade the controller. In all of the work described in this monograph, differentiation is avoided.

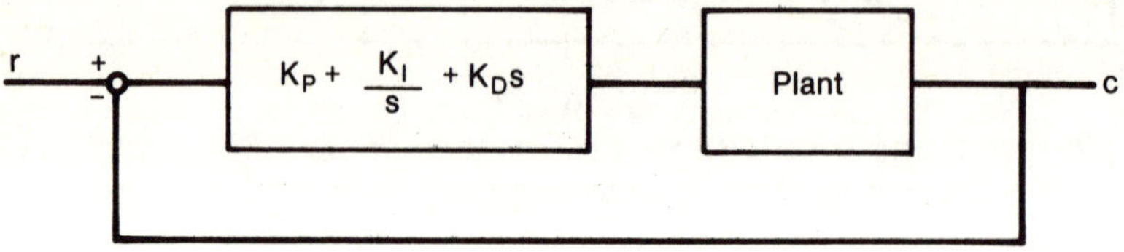

FIGURE 2-6
Plant with ordinary PID controller.

Reference

1. R. M. Phelan, *Automatic Control Systems*, Cornell University Press, Ithaca, NY, 1977.

CHAPTER III

Saturation Operation with Linear Feedback

In recent years automatic feedback control systems have played an increasingly important role in science and industrial technology. As more application opportunities are identified, more demands are placed upon the control system engineer to obtain high performance in terms of response and stability in very complex systems. These demands are driven by the greater use of industrial robots, by the present concern for increased productivity, by the need to control flexible structures and by the availability of ever faster microprocessors and minicomputers. New design techniques and control algorithms are being developed to meet these demands. Some of these techniques represent a combination of traditional control theory and modern control theory. These recent developments will be described in this chapter.

The control of flexible structures and flexible machine components has become one of the important control problems of the 1980's and can be expected to continue as an important problem for many years. The problems range in physical size from large space structures that do not have the rigidity of earth-based configurations, to machining operations where flexibility manifests itself as tool chatter or poor tolerance capability. The latter problem is a precision engineering problem. Where tolerances of a few millionths of an inch are important, small amplitude vibrations at the tool-workpiece interface are important. Also, surface finish quality is influenced adversely by these vibrations.

The problems with flexible structures follow from the use of high strength alloys, which allow high strains and large displacements when loaded to the allowable stress. For example, a 300,000 psi alloy with a Young's modulus of 30×10^6 psi will undergo a 1 percent strain at the allowable stress! A current graphic example is the industrial robot. In many applications a robot must be significantly faster than its human counterpart in order to be economically competitive. Moreover, the capital investment cannot be too large or the payback period will be too long and cannot be justified. The speed and the cost of the robot are competing considerations that must be weighed carefully. For example, the speed of response can be improved by using a larger servomotor, but this raises the cost of the robot. One option is to reduce inertia (weight/distribution) by designing to the allowable stress of the material used in the robot arm(s). Higher strength alloys may also be a viable option. In either case, the redesigned arm(s) will be more compliant and susceptible to low frequency vibrations. This could produce the undesired effect of increasing the response time because the robot arm and its payload could vibrate for some time at its terminus where an attempt is made to stop the assembly abruptly after a rapid traverse. We believe, however, that the very flexible structure whose motions are controlled by an automatic feedback control system will be a viable solution to this problem. A second option for improving the speed of response is to use servomotors or actuators that are operated at or near saturation most of the time. Bang-bang operation is usually guaranteed to produce a minimum time traverse. These options are described here.

Objectives

Perhaps the most dramatic illustration of a controlled structure is an outstretched human arm that supports a load.

Within the physical limits of the arm, a range of loads can be supported without external deflections. In this sense the arm is an inelastic structure. Certainly, many changes in the muscle structure take place in order to produce this effect. One of the long-term goals of this research effort is to design structures and components that emulate this inelastic behavior. Control algorithms described in this chapter are the first steps toward achieving this goal.

Another goal is related to servomotors or actuators that are used only seldom at their rated capacity. Linear systems generally do not allow actuators to be driven to saturation. This is a nonlinear phenomenon that has found relatively little application in complex control systems because available analytical techniques are not applicable for large systems. The advantages of using actuators driven to saturation are attractive. Smaller, lighter, and less expensive actuators can be used. This, in turn, can result in even lighter supporting structure, which, in the case of industrial robots, should permit further improvements in response.

Robots were used in the foregoing discussion because they are the subject of much active interest and because the problems described are readily recognized. In precision engineering, a parallel or analogous problem is the control of metal removal (machining) as it occurs. That is, the cutting tool will be subject to continuous and "real-time" control *during the machining process*. This is true in-process control.

Automatic Control

Classical control theory was developed in the 1940's and made extensive use of Laplace domain and frequency domain analyses. Design tools included the Bode diagram, the Nyquist diagram, and the root locus plot. These techniques served well for systems where all but one or two feedback gains were specified or where it was possible to cascade a series of high-pass networks. Even for two unspecified gains that could not be isolated from each other, the methods were cumbersome because the graphical plots required three-dimensional presentation. As systems became more complex with more feedback interaction and coupling, more design flexibility was required, and the traditional design tools could not be used effectively where three or more feedback gains were to be specified.

In the late 1950's, the time domain state space analyses of modern control theory supplemented the traditional analyses. As we mentioned in Chapter 1, the name, modern control theory, is misleading since state space analyses date back to the work of Poincare, Lyapunov, and Gibbs around the turn of the century. Nonetheless, additional tools became available for control system analysts and designers. The state space analysis is neat and compact. The vector-matrix presentation of the system differential equations is not only attractive but also calls attention to the fact that dynamical systems are built up of first-order and second-order systems. This contrasts with classical control theory where attention often was focused on the system characteristic equation, which may be of very high order.

Optimal Linear Feedback Controller

Linear control systems described by vector-matrix differential equations were a straightforward application for Pontryagin's Minimum Principle. For the optimal linear feedback controller, the index of performance (a quadratic form of the state variables and the control variables) is selected by the analyst on the basis of experience and immediate requirements. Feedback gains, associated with each state variable, are time dependent and are obtained from the solution to the nonlinear matrix Riccati differential equation. In some applications, only the steady-state solution to the matrix Riccati equation is required. Then the feedback gains are constant in time. Two disadvantages of the optimal linear feedback controller analysis are that the index of performance is judgmental and that the horizon or final time must be specified where the feedback gains are time dependent. A third disadvantage is that the controller does not adapt readily to arbitrary and unexpected disturbances. The judgmental character of the index of performance, however, can be turned to good advantage in cases where the IP to be minimized can be selected as:

$$\mathrm{IP} = \int_0^{t_f} (\dot{u})^2 dt \qquad (3\text{-}1)$$

where $\dot{u}$ is the control variable. Here, u is a scalar, but the argument extends readily to vector quantities. A minimum occurs when $\dot{u} \rightarrow 0$, and this can give a soft bang-bang control. This is a significant benefit because the actuator will be used at or near saturation much of the time. The inability to handle arbitrary disturbances effectively and the need for a specified horizon remain disadvantages, however.

Problem Statement

In the following development we consider a feedback control algorithm in the time domain with complete state variable feedback. The algorithm is an extension of compensation networks with integral and proportional compensation (1). Within a linear system framework, the problems are as follows:

1. Assign the poles to obtain a desired form of response in the time domain
2. Design an error compensation system that permits an actuator to be driven to saturation within a linear framework
3. Design a control system to control the angular velocity of a rotating inertia which is driven by a servomotor through an elastic member
4. Design a system to be insensitive to arbitrary disturbances

Solution Method

Illustrative examples will be used to show how these problems can be solved. We consider first a dc permanent-magnet servomotor with a completely linear error compensation network. The same motor is then considered with a piecewise linear compensation network to achieve torque and speed limiting. Finally, the power of pole assignment of the new control algorithm is illustrated with a load driven by a motor through a flexible or extensible belt.

DC SERVOMOTOR WITHOUT LIMITERS

A simple permanent-magnet dc servomotor shown in Figure 3-1 (1) is used in this section to show how to solve the first two problems listed in the Problem Statement. This motor is used to drive a "daisywheel" print element for a word processor. The dynamical equations that describe the motor are:

$$
\begin{aligned}
L_a \frac{di_a}{dt} &= -R_a i_a + e_a - e_b \\
e_b &= K_b \omega \\
T_m &= K_i i_a \\
J \frac{d\omega}{dt} &= -B\omega + T_m + T_D \\
\frac{d\theta}{dt} &= \omega
\end{aligned}
\qquad (3\text{-}2)
$$

where:

i_a = armature current, amperes

e_a = armature voltage supplied by dc amplifier, volts

R_a = armature resistance, 5 ohms

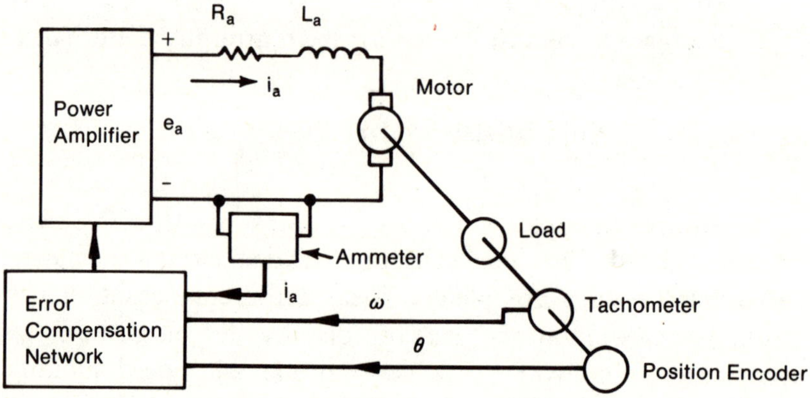

FIGURE 3-1
Simple dc servo motor with feedback control.

e_b = back emf, volts

K_b = back emf constant of motor, 0.02125 V/rad/sec

T_D = disturbance torque, in.-oz.

T_m = torque output of motor, in.-oz.

L_a = motor inductance, 0.005 Henry

K_i = torque constant of motor, 3 oz.-in./amp

J = mass polar moment of inertia of motor and wheel, 0.0015 oz.-in.-sec^2

B = viscous friction coefficient, 0.06 oz.-in.-sec

θ = angular position of motor, radians

ω = angular velocity of motor, rad/sec

The block diagram of the motor alone is shown in Figure 3-2. The state space model of the motor is neat and compact:

$$\begin{bmatrix} \dot{i}_a \\ \dot{\omega} \\ \dot{\theta} \end{bmatrix} = \begin{bmatrix} -R_a/L_a & -K_b/L_a & 0 \\ K_i/J & -B/J & 0 \\ 0 & 1 & 0 \end{bmatrix} \begin{bmatrix} i_a \\ \omega \\ \theta \end{bmatrix} + \begin{bmatrix} 1/L_a \\ 0 \\ 0 \end{bmatrix} e_a + \begin{bmatrix} 0 \\ 1/J \\ 0 \end{bmatrix} T_D \qquad (3\text{-}3)$$

The error compensation network makes use of complete state variable feedback, the integral of the output error, and the integral of each of the "errors" associated with the remaining state variables. An optional approach is the use of a single

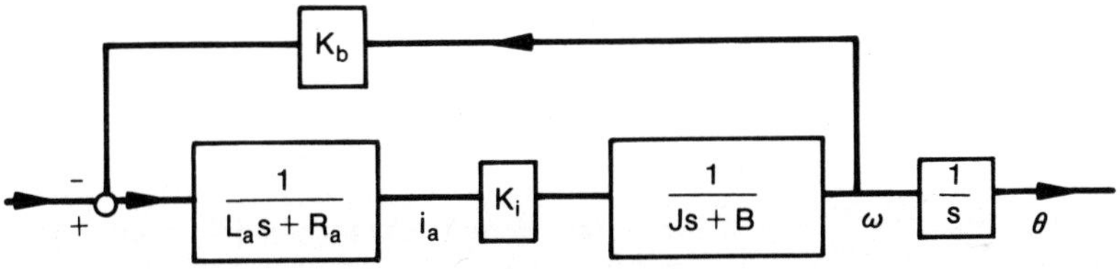

FIGURE 3-2
Block diagram of dc servo motor.

integrator for the output state variable. For convenience, we describe these systems as Integral Error with State Feedback (IESF). The complete block diagram with the error compensation network is shown in Figure 3-3. If we call the outputs of the three integrating amplifiers y_1, y_2, and y_3, the state space model of the entire system is

$$\dot{X} = \begin{bmatrix} \dot{y}_1 \\ \dot{y}_2 \\ \dot{y}_3 \\ \dot{i}_a \\ \dot{\omega} \\ \dot{\theta} \end{bmatrix} = \begin{bmatrix} 0 & 0 & 0 & 0 & 0 & -K_1 \\ K_2 & 0 & 0 & 0 & -K_2 & -K_{D1}K_2 \\ 0 & 1 & 0 & -1 & -K_{D2} & 0 \\ 0 & 0 & 1/L_a & (-R_a-K_{D3}/L_a & -K_b/L_a & 0 \\ 0 & 0 & 0 & K_i/J & -B/J & 0 \\ 0 & 0 & 0 & 0 & 1 & 0 \end{bmatrix} \begin{bmatrix} y_1 \\ y_2 \\ y_3 \\ i_a \\ \omega \\ \theta \end{bmatrix} + \begin{bmatrix} K_1\theta_i \\ 0 \\ 0 \\ 0 \\ 0 \\ 0 \end{bmatrix} + \begin{bmatrix} 0 \\ 0 \\ 0 \\ 0 \\ 1/J \\ 0 \end{bmatrix} T_D \tag{3-4}$$

The Laplace transform of the system response is:

$$\theta = \frac{K_i K_1 K_2 K_3}{JL_a D(s)} \theta_i + \frac{L_a s^4 + (K_{D3} + R_a)\, s^3 + K_3 s^2}{JL_a D(s)} T_D \tag{3-5}$$

where:

$$\begin{aligned} D(s) = {} & [L_a J s^6 + (L_a B + R_a J + JK_{D3}) s^5 \\ & + (R_a B + BK_{D3} + K_b K_i + JK_3) s^4 \\ & + (BK_3 + K_{D2} K_i K_3) s^3 + K_i K_2 K_3 s^2 \\ & + K_i K_{D1} K_2 K_3 s + K_i K_1 K_2 K_3] \,/\, L_a J \end{aligned} \tag{3-6}$$

From this transform we conclude:

1. The steady-state error for θ becomes zero for a step input for θ_i.
2. The steady-state effect on θ, of the disturbance torque T_D, will vanish for a constant or ramp disturbance. This

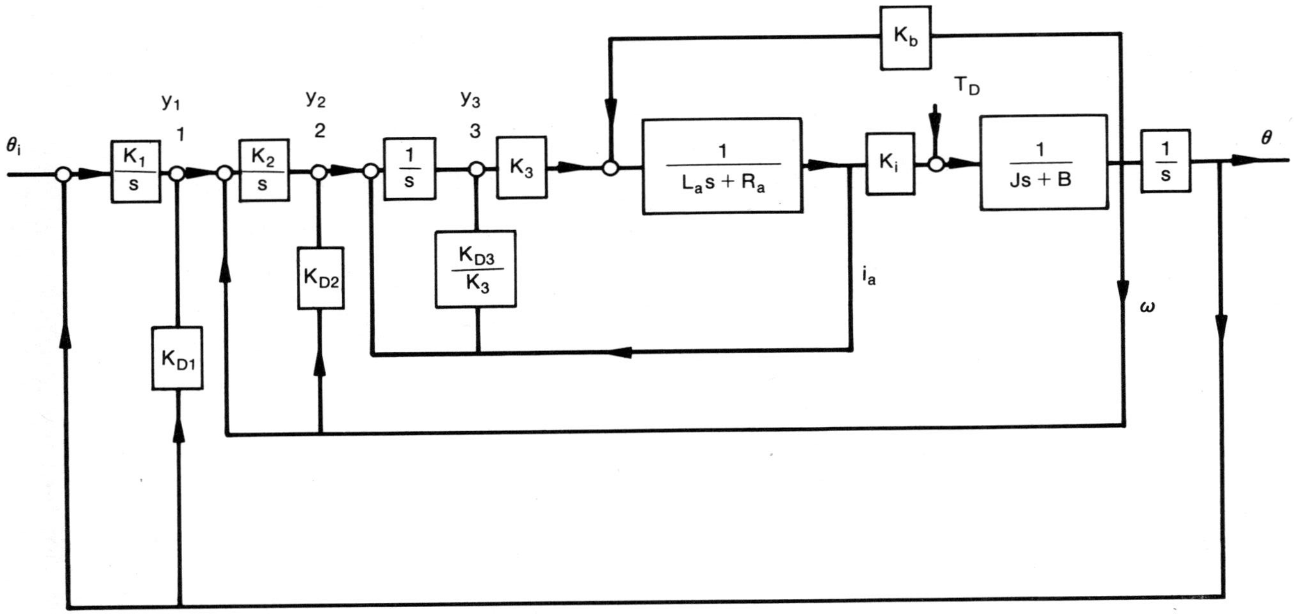

FIGURE 3-3
Block diagram of dc servo motor with error compensation, cascaded integrators, and state feedback.

assumes that the amplifier has enough capacity to keep up with a ramp disturbance. In any event, that is a remarkable property of the system.

3. The coefficients of $D(s)$ contain the unknown K_1, K_2, K_3, K_{D1}, K_{D2}, and K_{D3} in such a way that these six gains can be specified to place the roots of $D(s)$ anywhere in the s-plane. We have pole assignment capability. We can demonstrate this from Equation 3-6 by setting the right-hand side to zero

$$\begin{aligned} s^6 &+ (1/L_aJ)(L_aB + R_aJ + JK_{D3})s^5 \\ &+ (1/L_aJ)(R_aB + BK_{D3} + K_bK_i + JK_3)s^4 \\ &+ (1/L_aJ)(BK_3 + K_{D2}K_iK_3)s^3 + (1/L_aJ)(K_iK_2K_3)s^2 \\ &+ (1/L_aJ)(K_iK_{D1}K_2K_3)s + (1/L_aJ)(K_iK_1K_2K_3) = 0 \end{aligned} \tag{3-7}$$

If we wish to place the system poles at $s = -a$, where a is a real and positive number, we expand $(s + a)^6$ to obtain

$$(s + a)^6 = s^6 + 6as^5 + 15a^2s^4 + 20a^3s^3 + 15a^4s^2 + 6a^5s + a^6 \tag{3-8}$$

and then equate the coefficients of like powers of s in Equations 3-7 and 3-8. Six algebraic equations result that can be solved directly and sequentially for K_{D3}, K_3, K_{D2}, K_2, K_{D1}, and K_1.

At this point, we have accomplished the goals of pole placement and lack of sensitivity to certain arbitrary disturbances. We note that these goals could have been achieved, but to a lesser degree, if the second and third integrators are removed or bypassed. The insensitivity to external disturbances is reduced. Only a step disturbance can be handled effectively. The block diagram for this case, Figure 3-4, shows that this approach uses state feedback with a single integral compensation of the error signal.

The transform for this system is:

$$\theta = \frac{K_1}{D(s)}\,\theta_i + \frac{s(K_4 + R_a + L_as)}{K_iD(s)}\,T_D \tag{3-9}$$

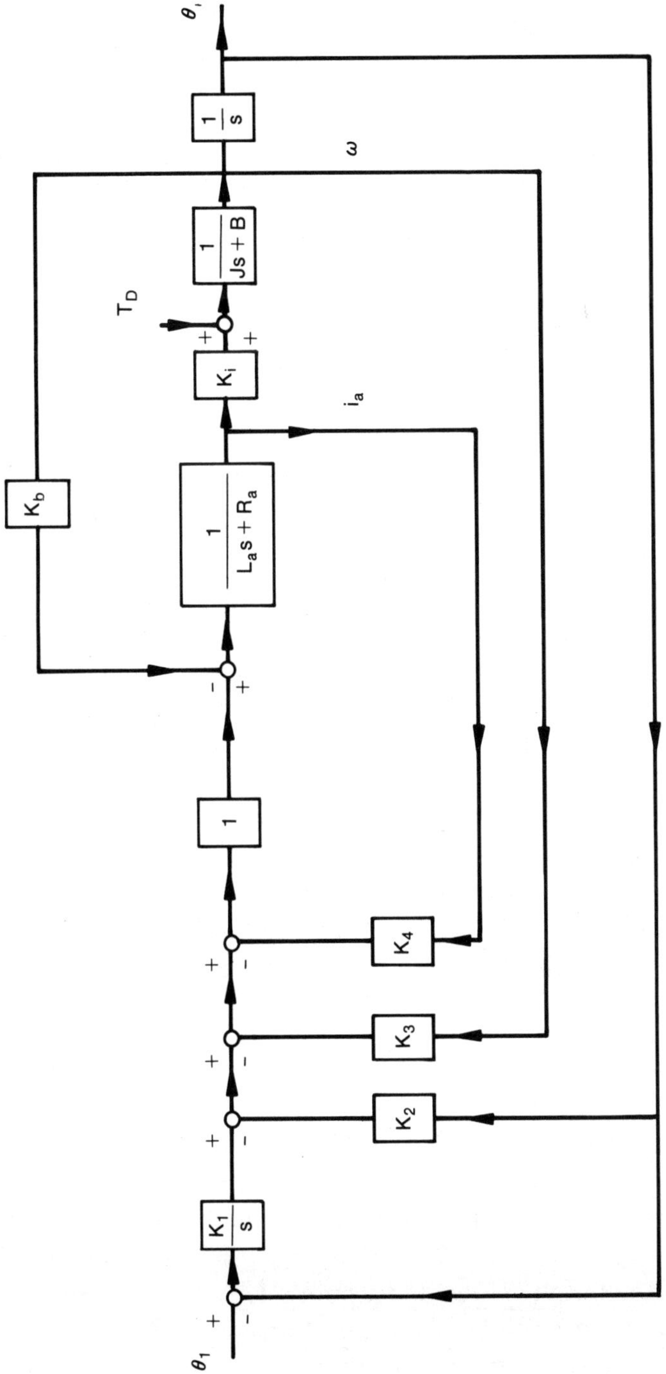

FIGURE 3-4
Block diagram of dc servo motor with one stage of integral compensation and state feedback.

where

$$D(s) = K_1 + sK_2 + s^2\left(K_3 + \frac{BK_4}{K_i} + K_b + \frac{R_aB}{K_i}\right) + \frac{s^3}{K_i}(K_4J + R_aJ + L_aB) + s^4\frac{L_aJ}{K_i}$$

The order of the system has been reduced by two. It will show zero steady-state error to a step torque disturbance but not to the ramp disturbance as in the previous system.

For the system of Figure 3-3, whose transfer function is given in Equation 3-5, repeated roots were selected at -100. The unknown gains are computed to be:

$$\begin{aligned} K_1 &= 666.67 & K_{D1} &= 40.0 \\ K_2 &= 6.2972 & K_{D2} &= 0.063963 \\ K_3 &= 595.50 & K_{D3} &= -2.20 \end{aligned} \qquad (3\text{-}10)$$

The high gains for K_1 and K_3 follow from assigning the closed-loop poles at -100 to provide good time response. A consequence of this is that physical implementation will require serious consideration of a direct digital controller (DDC) with limiters placed at the outputs of one or more of the summing points in the compensation network.

For a unit step input at θ_i, the response $\theta(t)$ is "deadbeat," i.e., absolutely no overshoot and no oscillatory behavior. The output $\theta(t)$ is shown in Figure 3-5 as the response without limiters.

DC SERVOMOTOR DRIVEN TO SATURATION

The second specification listed in the problem statement is the operation of a dc motor at saturation within a linear framework. Response time should be improved substantially. This can be done by moving the closed-loop poles to the left of -100 and then limiting the output of summing points 1, 2, and 3 of Figure 3-3. Note that point 1 can be associated with the input signal for the angular velocity ω, point 2 with the current i_a, which is directly proportional to the torque, and point 3 with the

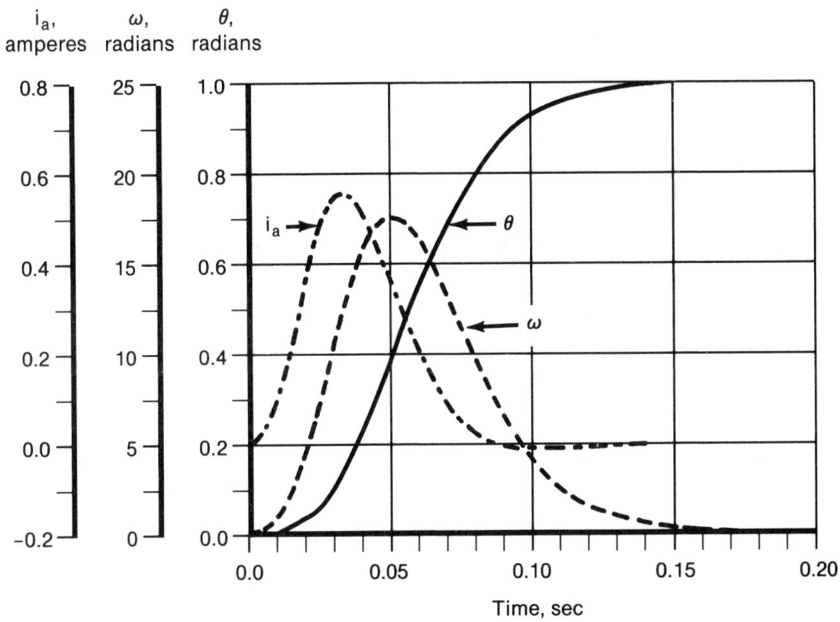

FIGURE 3-5
Response of dc servo motor to unit step input — no constraints.

armature voltage. The input to an integrator, whose output is at a limit or constraint, is restricted so that it cannot have the same sign as the output of that integrator. This prevents a situation known as "reset windup". This occurs in the time response of limited I-controllers to a large step input or a change in the reference (set) point. For a positive error, for example, the error is integrated over time and continues to increase beyond the constraint into the saturation region. Even as the error returns to the linear range, the integral of the positive error continues to accumulate until the sign changes. The integrator cannot "unload" until the error changes sign, thus keeping the controller output at or near the saturation value even though the system is near the new set point. This nonlinear phenomenom can cause large and repeated overshoots in the system response, and it can cause instability. The algorithm for preventing reset windup of a constrained integrator is shown schematically in Figure 3-6. Consider now point 1 in Figure 3-3. When the output of this point is at a constraint value, the control system may be considered to be of reduced order (four instead of six) where the

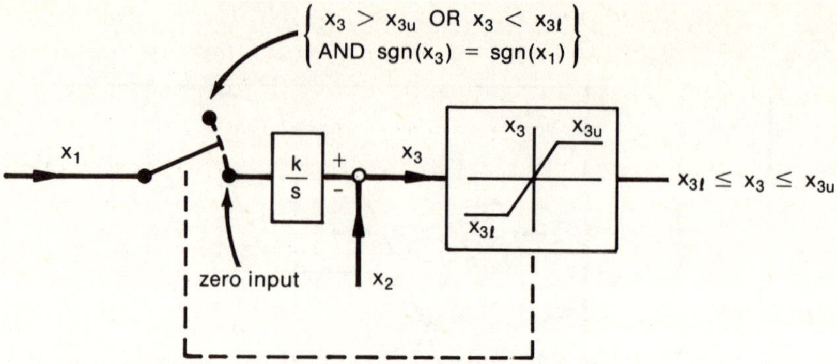

FIGURE 3-6
Schematic of logic sequence for limited integrator.

controlled variable is the angular velocity ω. Hence, it will be necessary to schedule or recompute the gains and the pole locations accordingly. The gain K_{D3} is left unchanged so that the input to the power amplifier is continuous. Consider next point 2. When the output of this point is at a limiting value, the order of system is "reduced" from four to two. The controlled variable may be regarded as the current i_a, which is directly proportional to the torque. Finally, when the output of point 3 is limited, the constrained variable is the output e_a of the power amplifier. This output voltage is the control variable.

Effective constraints are achieved by limiting the output of summing points 1, 2, and 3. This is equivalent to placing limiters or constraints on the state variables and the control variable. Computational experience showed that the ordering of hierarchy of the variables was important in implementing the limits. Thus, a limit on the angular velocity ω could be imposed alone. A limit on the current i_a, however, required a limit on the angular velocity to prevent a form of reset windup at the input to the second integrator. Numerical simulation will show whether it is necessary to constrain e_a. As more experience is gained with these limiting techniques, new or modified strategies may be found. An effective approach for selecting the limiting values for these three points is based upon the variables ω, i_a, and e_a for some reference operating condition. We selected the unconstrained condition for which the maximum current is 0.5505 amp. This value was used as a limit or allowable current for three other sets of operating conditions that are described in the following. We want to maximize the amount of time we can hold the motor at or near the maximum allowable current. This will result in maximum torque over a significant part of the cycle, which will reduce greatly the time required for the motor to traverse one radian. In order to cause the servomotor to be driven to saturation current during a substantial part of the cycle, we relocate the system's poles to the left of the reference condition at $s = -100$. If no constraints were placed on the system at this point, note that the motor would respond much faster and the armature current would be significantly higher than the reference condition. In order to prevent this from happening, we limit the output of point 2 in Figure 3-3, so that the value of the output of that point is 0.5505 amp. Experience with numerical solutions has shown that a similar constraint must be placed on the output of point 1 in Figure 3-3 to prevent a form of reset windup. The value of this constraint, which we can associate with the angular velocity, must be determined through an iterative trial and error process. A good trial value for the angular velocity is the steady-state value associated with a current of 0.5505 amp. A calculation based on the system equations will show that this angular velocity is about 27.5 radians per second. Experience with numerical solutions for the system show, however, that the appropriate limiting value of the angular velocity may be somewhat less than 27.5. The best values were between 22 and 23 radians per second. When the constraint value was greater than 23 radians per second, the system experienced overshoot towards the final part of the traverse and the time required to settle out at the steady-state value was substantially increased. We see then that the constraint on the angular velocity plays the role of an anticipatory constraint on the current. For the first case, we specify repeated poles at $s = -200$. The feedback gains and the integrator gains are computed to be:

$$\begin{aligned} K_1 &= 2667 & K_{D2} &= 0.1267 \\ K_{D1} &= 80 & K_3 &= 2725.5 \\ K_2 &= 22.01 & K_{D3} &= 0.8 \end{aligned} \tag{3-11}$$

When the output of only point 1 in Figure 3-3 is limited, K_{D3} is unchanged at 0.8 but the other gains are:

$$K_2 = 8.3488$$
$$K_{D2} = 0.091317 \qquad (3\text{-}12)$$
$$K_3 = 2425.5$$

For this condition, the repeated poles are found at $s = -300$. When the output of point 2 in Figure 3-3 is limited, K_{D3} remains unchanged again at 0.8 but $K_3 = 1682$ and the repeated poles are at $s = -580$. The results of the numerical simulation for this case are shown in Figure 3-7 where it is seen that the response time is reduced by about one-third from the case shown in Figure 3-5.

For Case II, repeated poles are specified at $s = -300$. The various gains are:

$$K_1 = 6000 \qquad K_{D2} = 0.1924$$
$$K_{D1} = 120 \qquad K_3 = 6355.5 \qquad (3\text{-}13)$$
$$K_2 = 47.79 \qquad K_{D3} = 3.8$$

When the output of point 1 in Figure 3-3 is limited, K_{D3} remains unchanged at 3.8, but the other gains are:

$$K_2 = 18.05$$
$$K_{D2} = 0.1404 \qquad (3\text{-}14)$$
$$K_3 = 5680.5$$

The new repeated poles are at $s = -450$. When the output of point 2 in Figure 3-3 is limited, K_{D3} remains unchanged at 3.8 but $K_3 = 3872$. The two repeated poles are at $s = -880$. The results of the numerical simulation for this case are shown in Figure 3-8.

For Case III, we specify repeated poles at $s = -350$ and the resulting gains are:

$$K_1 = 8167 \qquad K_{D2} = 0.2255$$
$$K_{D1} = 140.0 \qquad K_3 = 8733 \qquad (3\text{-}15)$$
$$K_2 = 64.44 \qquad K_{D3} = 5.3$$

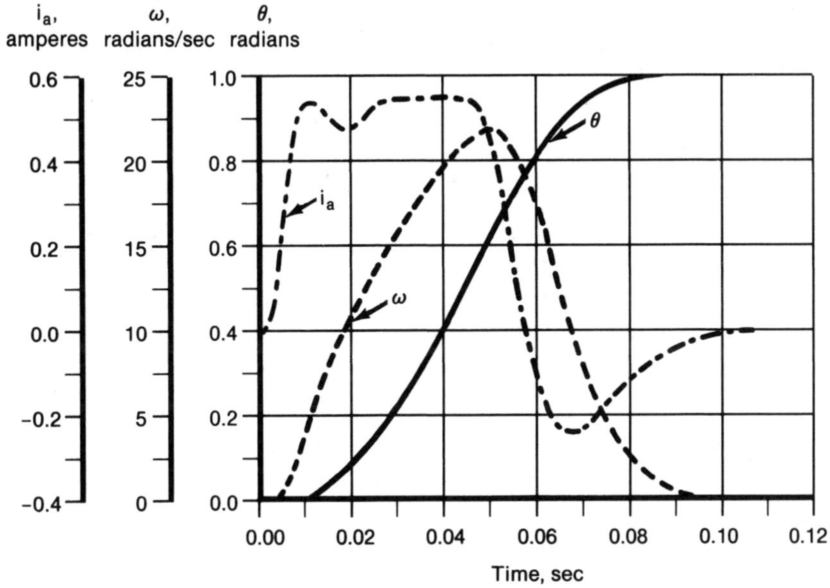

FIGURE 3-7
Response of constrained servo motor to unit step input; specified repeated poles at s = -200.

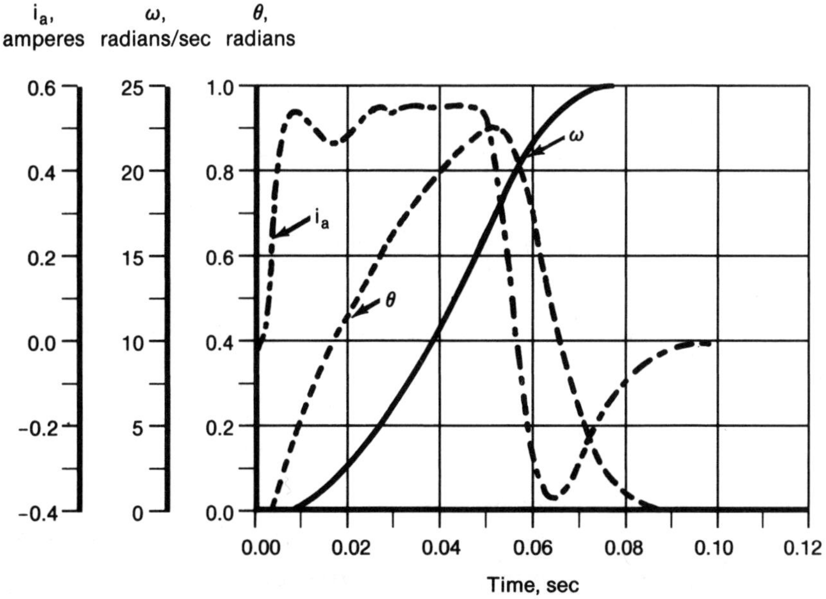

FIGURE 3-8
Response of constrained servo motor to unit step input; specified repeated poles at s = -300.

When the output of point 1 in Figure 3-3 is limited, K_{D3} is unchanged at 5.3, but the other gains are:

$$\begin{aligned} K_2 &= 24.30 \\ K_{D2} &= 0.1652 \\ K_3 &= 7814 \end{aligned} \tag{3-16}$$

The four repeated poles are at $s = -525$. Finally, when the output of point 2 in Figure 3-3 is limited, K_{D3} is again unchanged at 5.3, and there are two repeated poles at $s = -1030$. The results of numerical simulations for this case are shown in Figure 3-9. There it is seen that the response time is reduced almost by one-half when compared with the original situation shown in Figure 3-5. It can be seen also that the motor certainly does operate at or near saturation for a significant part of this cycle.

When repeated poles are specified to the left of $s = -350$, it was found that there is considerable overshoot bcyond $\theta = 1.0$ so that we no longer achieve the required deadbeat response. Moreover, the time required for the system to settle out is

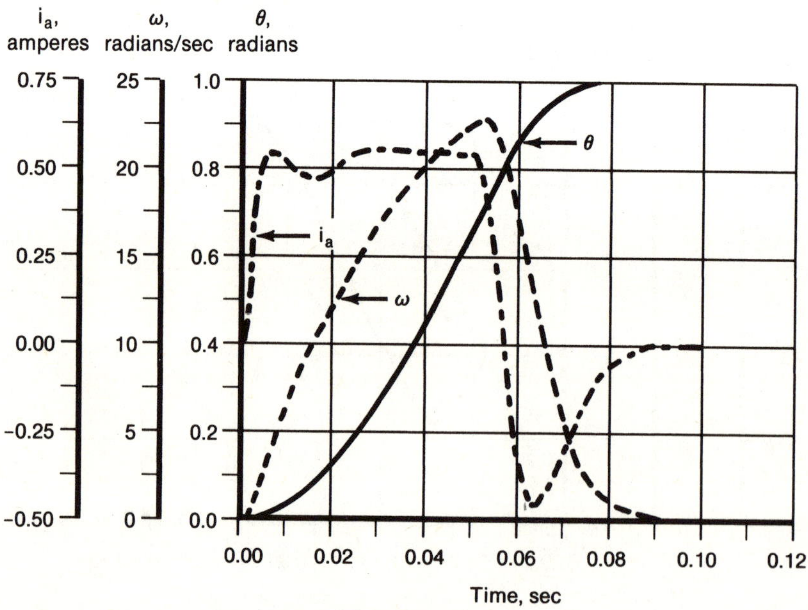

FIGURE 3-9
Response of constrained servo motor to unit step input; specified repeated poles at s = -350.

increased significantly so that the assignment of the specified roots beyond $s = -350$ is not to our advantage.

The three cases shown in Figures 3-7, 3-8, and 3-9 are characterized by an almost linear increase in the angular velocity followed by an almost linear decrease in the angular velocity. There are other situtations where the angular velocity exhibits a relatively flat behavior between the rise and decrease regions. One such case is shown in Figure 3-10. Here, the step input is 6 radians rather than 1 radian, and we see that the angular velocity is at or near the limit value of 27.525 radians/second corresponding to the limited current of 0.5505 amp. It was found that the specified poles could not be to the left of $s = -250$ because the response was not deadbeat.

It is interesting to note in all of the curves shown that the current, which is limited to some specified value, exhibits a sharp rise followed by a slight dip and then a continued rise back to the limiting value. Numerical simulations showed that the dip occurred at the same time that the output of point 3 in Figure 3-3 approached the limiting angular velocity value. It is emphasized that the best or optimum value of the assigned poles

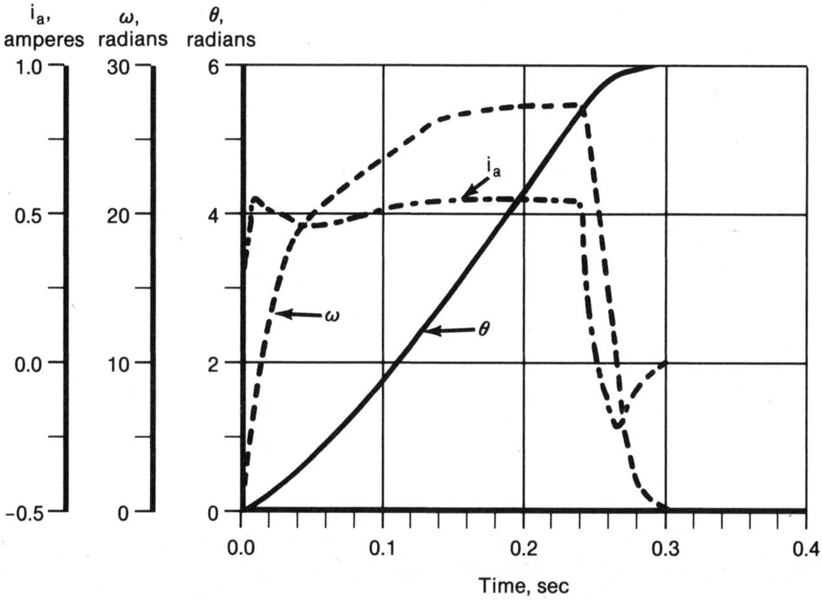

FIGURE 3-10
Response of constrained servo motor to unit step input of six radians; specified repeated poles at s = -250.

and the appropriate limits on the state variables are determined most quickly by carrying out simulations on the digital computer. The author's FORTRAN listing, called CONTROL, is in the Appendix.

This development has been concerned with a simple dc servomotor driving an inertia load rigidly connected to the motor. Elastic effects were neglected so that attention could be focused on (1) the pole allocation provisions of IESF compensation, and (2) the ability to operate actuators at or near saturation current.

ELASTIC SYSTEMS

We turn our attention to a simple computer tape drive system where the load is driven by a permanent magnet dc servomotor through an elastic drive belt (Figure 3-11 (1)). The objective is to design a feedback control system so that we can change the speed of the load from one value to another in some reasonable manner. Specifically, we seek $\omega_L(t)$ for a step input to the system. Overshoot is to be avoided if possible. We do not consider limiters or constraints so that we can concentrate on the problem of motion control through a spring. The system equations are:

$$\begin{aligned} J_m\ddot{\theta}_m &= -K_e\,\dot{\theta}_m - K_L\,(\theta_m - \theta_L) + T_m \\ J_L\ddot{\theta}_L &= -B_K\dot{\theta}_L - K_L\,(\theta_L - \theta_m) + T_D \\ T_m &= K_a i_a \\ e_a &= i_a R_a + K_b i_a \\ K_e &= K_b K_a / R_a + B_m \end{aligned} \tag{3-17}$$

where

$K_a/R_a = 36$ oz.-in.

$K_e = 6.92$ oz.-in./rad/sec

$B_L = 10$ oz.-in.-sec

$J_m = 0.023$ oz.-in.-sec^2

$K_L = 2857.6$ oz.-in./rad

$J_L = 7.24$ oz.-in.-sec^2

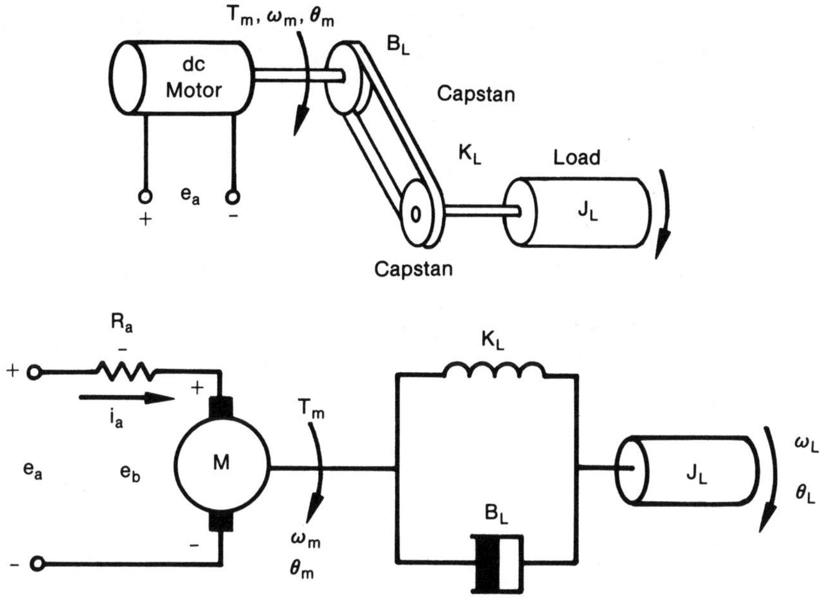

FIGURE 3-11
Tape drive motor system.

and

B_L = damping constant of load

B_m = damping constant of motor

e_a = output voltage of amplifier

i_a = armature current

J_L = polar moment of inertia of load

J_m = polar moment of inertia of motor

K_b = back-emf constant

R_a = armature resistance

T_D = torque disturbance

θ_L = angular position of load

θ_m = angular position of motor

ω_L = angular velocity of load

ω_m = angular velocity of motor

The state-space model for the motor-load combination with an IESF controller is

$$\begin{bmatrix} \dot{y}_1 \\ \dot{y}_2 \\ \dot{y}_3 \\ \dot{\omega}_m \\ \Delta\dot{\omega} \\ \dot{\omega}_L \\ \Delta\dot{\theta} \end{bmatrix} = \begin{bmatrix} 0 & 0 & 0 & 0 & 0 & -K_1 & 0 \\ K_2 & 0 & 0 & 0 & 0 & -K_2 K_{D1} & -K_2 \\ 0 & K_3 & 0 & 0 & -K_3 & 0 & -K_3 K_{D2} \\ 0 & 0 & 1/J_m & -K_e/J_m & -K_{D3}/J_m & 0 & K_L/J_m \\ 0 & 0 & -1/J_m & K_e/J_m & K_{D3}/J_m & -B_L/J_L & -K_L(1/J_L+1/J_m) \\ 0 & 0 & 0 & 0 & 0 & -B_L J_L & -K_L/J_L \\ 0 & 0 & 0 & 0 & 1 & 0 & 0 \end{bmatrix} \begin{bmatrix} y_1 \\ y_2 \\ y_3 \\ \omega_m \\ \Delta\omega \\ \omega_L \\ \Delta\theta \end{bmatrix} + \begin{bmatrix} K_1 \Omega_L \\ 0 \\ 0 \\ 0 \\ 0 \\ 0 \\ 0 \end{bmatrix} + \begin{bmatrix} 0 \\ 0 \\ 0 \\ 0 \\ 1/J_L \\ 1/J_L \\ 0 \end{bmatrix} T_D$$

The block diagram is shown in Figure 3-12. The Laplace transform for the *angular velocity* of the load is:

$$\omega_L(s) = \frac{K_1 K_2 K_3 K_L s}{D(s)} \Omega_L(s) + \frac{J_m s^6 + (K_e - K_{D3}) s^5 + (K_L - K_3) s^4 - K_3 K_{D2} s^3 - K_2 K_3 s^2}{D(s)} T_D \tag{3-18}$$

where the system characteristic polynomial is:

$$\begin{aligned} D(s) = {} & J_L J_m s^7 + (J_L K_e + B_L J_m - J_L K_{D3}) s^6 \\ & + [(J_m + J_L) K_L + B_L K_e - B_L K_{D3} - J_L K_3] s^5 \\ & + [(K_e + B_L) K_L - J_L K_3 K_{D2} - B_L K_3] s^4 \\ & + (-B_L K_3 K_{D2} - J_L K_2 K_3) s^3 \\ & + (K_2 K_3 K_{D1} K_L - B_L K_2 K_3) s^2 + K_1 K_2 K_3 K_L s \end{aligned}$$

We can draw the following conclusions:

1. The steady-state error in ω_L vanishes for a step input at Ω_L.
2. The roots of the characteristic equation can be allocated to satisfy design requirements.

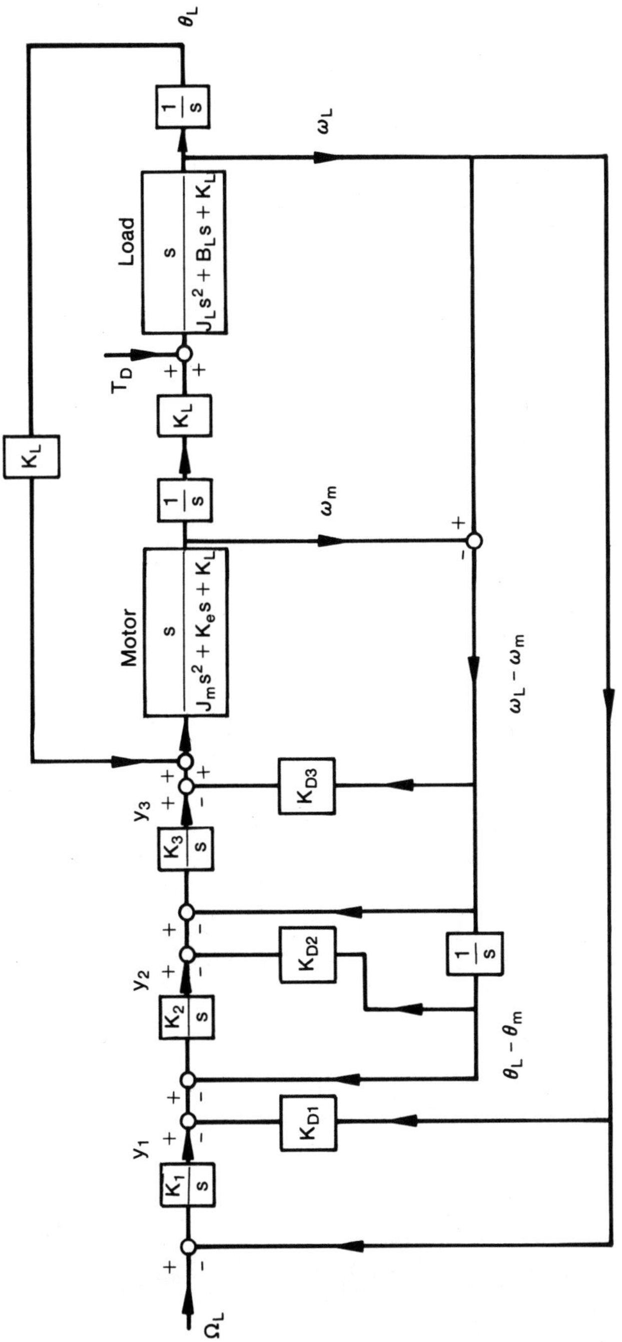

FIGURE 3-12
Block diagram of tape drive motor system with cascaded error compensation and state feedback.

3. There will be zero steady-state error in ω_L for a step disturbance torque, T_D.

Figure 3-13 shows the time response for repeated real roots at $s = -100$. The feedback gains are:

$$
\begin{aligned}
K_1 &= 1.720 \qquad & K_{D1} &= -9.971 \times 10^{-2} \\
K_2 &= 60029 \qquad & K_{D2} &= 801.0 \\
K_3 &= -564.3 \qquad & K_{D3} &= 5.3
\end{aligned}
\tag{3-19}
$$

ROBUSTNESS

Some indication of the robustness of these systems is obtained by considering the tape drive system. Robustness is generally considered to be the ability of a system to perform well

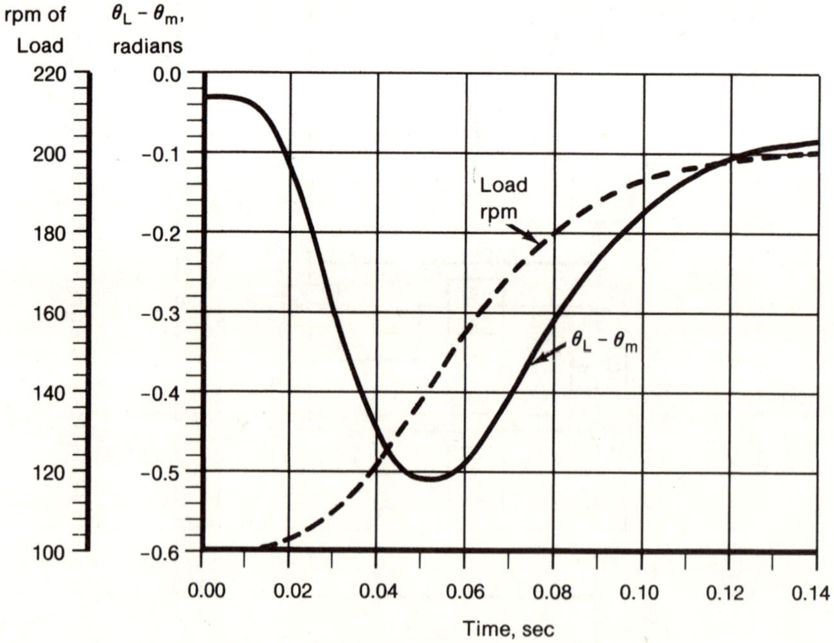

FIGURE 3-13
Response of tape drive system to step input of 100 rpm; initial speed is 100 rpm.

in the face of (1) substantial changes in system parameters or (2) substantial external disturbances.

The relative lack of sensitivity to changes in the driven load, J_L, is shown in Figures 3-14 and 3-15, where the nominal load J_L has been divided by $\sqrt{3}$ and multiplied by $\sqrt{3}$, respectively. Note that this is a total range of 3:1 for the driven load. All other parameters and gains remained the same. If the load were expected to change by this amount, a more appropriate approach would be to provide gain scheduling to accommodate the given load and still retain deadbeat response.

The response to an external torque, equal and opposite to the maximum torque in Figure 3-11, is shown in Figure 3-16. The speed drops about 50 rpm but returns to the desired value in about 0.15 seconds without any overshoot.

The author's FORTRAN listing for this simulation appears in the Appendix as CAPSTAN.

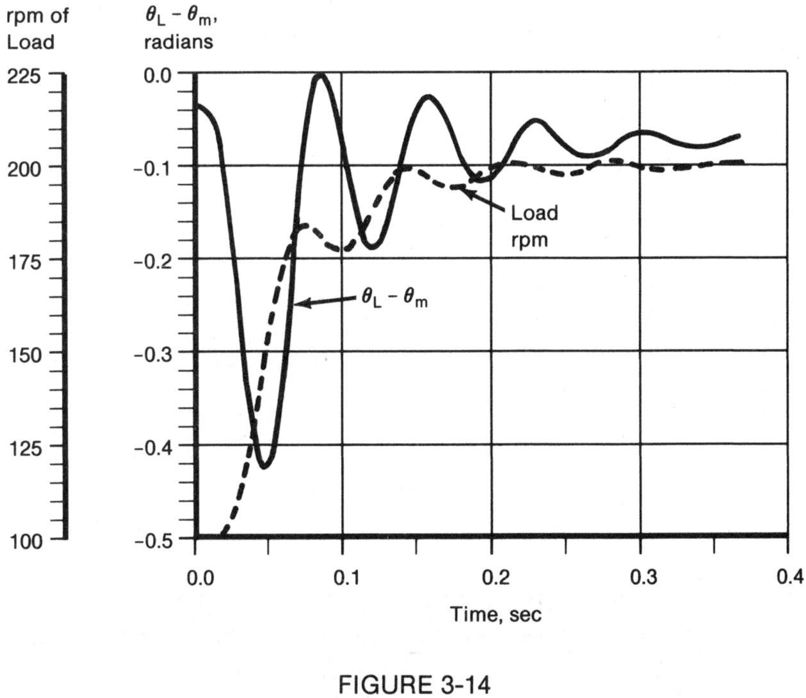

FIGURE 3-14
Response of tape drive system for J_L = 4.18 oz-in.-sec^2; feedback gains same as for Figure 3-12.

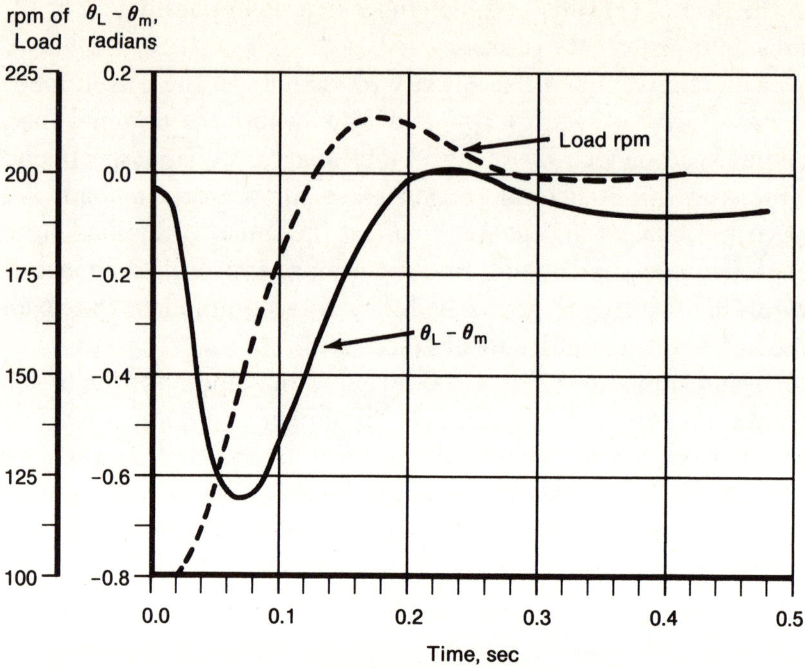

FIGURE 3-15
Response of Tape Drive System for J_L = 12.54 oz-in.-sec^2; feedback gains same as for Figure 3-12.

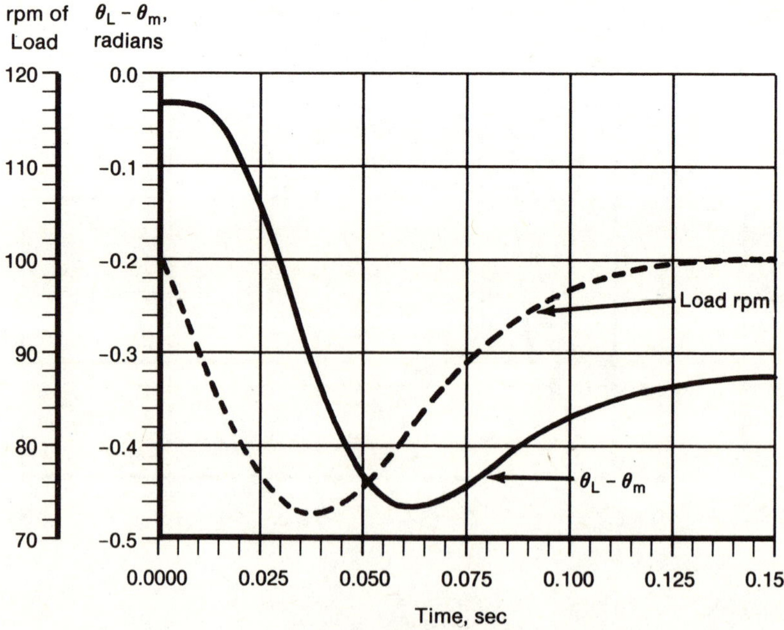

FIGURE 3-16
Response of tape drive system to step disturbance, T_D = 811 oz-in.

Discussion and Conclusion

We have described error compensation alogorithms for feedback control systems that permit design in the time domain. The algorithms, a combination of state feedback and integral error compensation, are an extension of techniques used in both classical control theory and modern control theory. In some respects, the algorithms may be considered a traditional implementation of linear optimal control systems with state variable and control variable constraint capability. We have shown how to:

1. Obtain a desired time response with pole placement ability
2. Design robust control systems
3. Operate actuators at or near saturation for a significant part of the cycle within a linear framework
4. Design controllers for systems with elastic compliance

The algebraic operations required to obtain the various transfer functions were facilitated greatly by the use of REDUCE™ and MACSYMA™.

Reference

1. Kuo, B., *Automatic Control Systems*, Fourth Edition, Prentice-Hall, Inc., Englewood Cliffs, New Jersey, 1982.

CHAPTER IV

Interaction and Noninteraction in Multi-Input, Multi-Output Systems

Introduction

Precision machining operations with real-time gaging and control, automatic control of gas turbines, and in-process control of chemical processes are representatives of important classes of feedback control systems that are characterized by more than one input and more than one output, i.e., multivariable systems with multi-inputs and multi-outputs. Such systems are the subject of many recent and current investigations (see References 1 through 11 at the end of this chapter). Often, there are difficulties with the accessibility of the output states, and these problems must be addressed effectively with observers to realize a satisfactory feedback control system. Even where it is possible to observe all of the state variables, there remains the problem of selecting feedback gains to achieve some desired performance level. Hence, there is an immediate concern about the nature of the time response. In multivariable systems with significant coupling, one performance criterion is the lack of interaction among the output variables when it is required to change only one of them. For example, in gas turbine applications it may be required to change shaft speed while maintaining certain pressures and temperatures within specified limits. Moreover, it is usually desirable to keep these limits very small. One approach to eliminate interaction is to diagonalize the plant transfer function matrix with a decoupling filter and then use a Proportional-Integral-Derivative (PID) controller. A second approach, superior to the first, uses complete state feedback with integral compensation of the errors but without a decoupling filter. That approach is used in this present chapter. It is referred to as Integral Error plus State Feedback (IESF) and is a generalization of the pseudo-derivative feedback controller (12).

A comparison is made of the performance of these two feedback control strategies (PID vs. IESF) for use with multivariable systems to reduce interaction or coupling effects. Comparisons are made for the proportional-integral-derivative controller with and without decoupling filters. Advantages of decoupling filters are the ability to control one output with one input, and the ability to apply the standard analytical techniques that are available for the study of single-input/single-output systems. Disadvantages of decoupling filters include:

1. The requirement to know the plant parameters accurately since incomplete decoupling may occur and this could reduce disturbance attenuation (13)
2. Possible loss of controllability due to pole cancellation
3. Possible physical unrealizability of the decoupling filter (11)

It is demonstrated that decoupling filters are not required for the integral error plus state feedback controller; hence, no comparisons were made for IESF control with such filters.

We consider a specific 2×2 (two inputs, two outputs) plant with dynamic coupling that results in a sixth-order system. It is simulated numerically with CSMP (Continuous Systems Modeling Program) (14) to determine the time response to a step input. While the two control strategies are substantially different, every effort is made to compare them equitably. Feedback gains for the proportional-integral-derivative controller are determined with the aid of the Ziegler-Nichols rules (15) and also through the use of the Integral of Absolute value of Error (IAE) criterion. Feedback gains for the integral error plus state feedback method followed from pole assignment at $s = -25$. Time domain performance characteristics for the controllers are obtained.

The robustness of the two strategies is investigated by changing system time constants 25% and, where appropriate, leaving unchanged the decoupling filter to produce a worst-case scenario. The performance of the integral error plus state feedback controller, without a decoupling filter, is found to be significantly superior to the proportional-integral-derivative controller, with or without the decoupling filter, in reducing or eliminating interaction.

Decoupling Filters with Proportional-Integral-Derivative Control

A two-input/two-output multivariable system can be represented in block diagram form as shown in Figure 4-1.

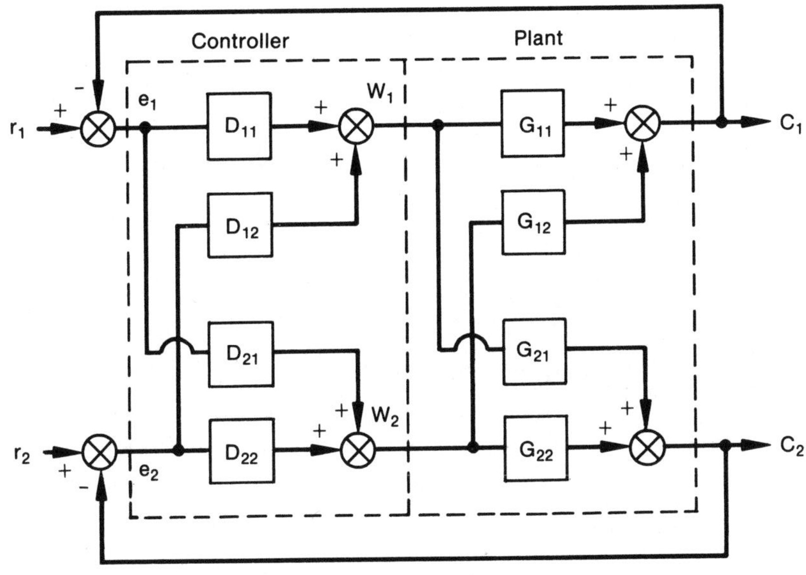

FIGURE 4-1
Two-input/two-output plant and controller with coupling.

"Design of Noninteracting Multivariable Feedback Control Systems Without Decoupling Filters". D. A. Baker and C. J. Maday, Transactions ASME, *Journal of Engineering for Gas Turbines and Power*, Vol. 107, No. 4, pp 845–850, October 1985.

In matrix notation, the plant is represented as

$$\{c\}_{2\text{x}1} = \{G\}_{2\text{x}2}\,\{w\}_{2\text{x}1} \tag{4-1}$$

and the controller as

$$\{w\}_{2\text{x}1} = \{D\}_{2\text{x}2}\,\{e\}_{2\text{x}1} \tag{4-2}$$

where

$$\{e\} = \{r\} - \{c\} \tag{4-3}$$

Combining the above equations and using unity feedback, the overall transfer function becomes (11)

$$[TF] = [I + [G]\,[D]]^{-1}\,[G]\,[D], \tag{4-4}$$

and

$$\{c\} = [TF]\,\{r\} \tag{4-5}$$

The forward transfer function of the control and plant together is

$$\{c\} = [G]\,[D]\,\{e\} \tag{4-6}$$

When the system is decoupled, the matrix

$$[G]\,[D] = \begin{bmatrix} G_{11}D_{11} + G_{12}D_{21} & G_{11}D_{12} + G_{12}D_{22} \\ G_{21}D_{11} + G_{22}D_{21} & G_{21}D_{12} + G_{22}D_{22} \end{bmatrix} \tag{4-7}$$

will be diagonal. Thus, by setting the off-diagonal terms equal to zero,

$$D_{21} = -\frac{G_{21}D_{11}}{G_{22}} \tag{4-8}$$

and

$$D_{12} = -\frac{G_{12}D_{22}}{G_{11}} \tag{4-9}$$

Substituting these into the main diagonal terms of Equation 4-7 yields,

$$[G][D] = \begin{bmatrix} \dfrac{[G]D_{11}}{G_{22}} & 0 \\ 0 & \dfrac{[G]D_{22}}{G_{11}} \end{bmatrix} \tag{4-10}$$

which represents two single-input/single-output forward transfer functions. Each of these resulting transfer functions can be treated separately at this point. One approach to feedback control is the use of proportional-integral-derivative compensation such as shown in Figure 4-2. The transfer function of the complete system is

$$\frac{c}{r} = \frac{K_c(1+T_I s + T_I T_D s^2)G_p(s)}{T_I s + K_c(1+T_I s + T_I T_D s^2)G_p(s)} \tag{4-11}$$

We note that this control introduces zeros in the transfer function. This contributes to fast response but also may increase

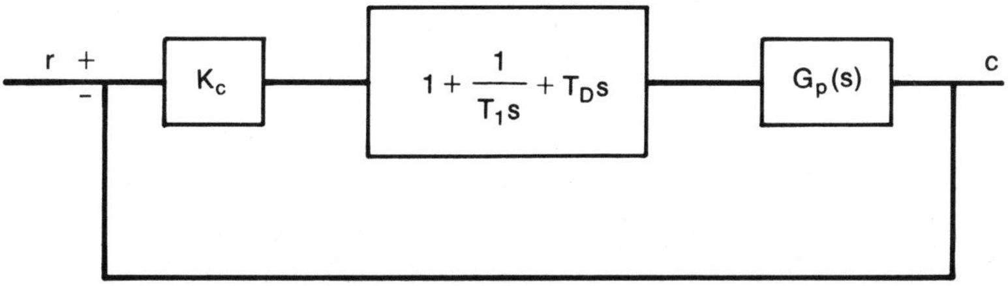

FIGURE 4-2
Single-input/single-output plant with PID (proportional-integral-derivative) control.

the oscillatory behavior of the system and the settling time. The three gains are selected with the aid of the Ziegler-Nichols rules,

$$K_c = \frac{1.2}{RL} \qquad T_I = 2L \qquad T_D = 0.5L \tag{4-12}$$

where R is the maximum slope of a tangent drawn to the unit-step input response of the plant to be controlled, and L is the time at which the tangent intersects the time axis.

Feedback gains are selected also to minimize the integrated absolute value of the error (IAE) to provide another measure of system performance. In general, the gains required to minimize this criterion gave better system performance than those obtained from the Ziegler-Nichols Rules.

Integral Error Plus State Feedback Control

The integral error plus state feedback controller is shown in Figure 4-3 for a single-input/single-output system (12). The gain $K_D(s)$ may contain derivatives of the output depending upon the order of $G_p(s)$ so that

$$K_D(s) = K_{D1} + K_{D2}s + \ldots\ldots + K_{Dn}s^{n-1} \tag{4-13}$$

where n is the order of the plant $G_p(s)$. For an nth order plant there are more than n viable versions of the integral error plus

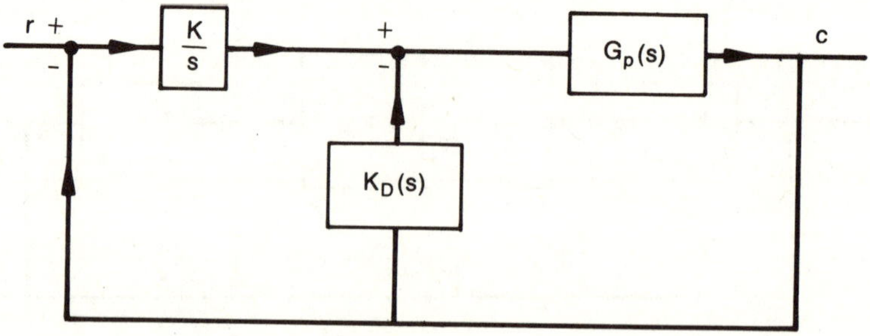

FIGURE 4-3
Single input/single output plant with IESF (Integral error plus state feedback) control.

state feedback that permit pole assignment. One version with a single-stage integration is shown in Figure 4-3. We note that a similar system could use complete state variable feedback rather than successive derivatives of the output. This would be the preferred approach because it eliminates the computational problems associated with determining derivatives. A variation of this system, without the integrator but with complete state feedback, gives the optimal linear regulator or servomechanism. Other versions, also referred to as subvariable control (16), include cascaded integrators such as shown in Figure 4-4. As more integrators are added, the feedback gains generally must be increased to maintain system response. Sufficiently high gains virtually necessitate the use of direct digital controllers rather than analog devices for satisfactory performance. These are treated in Chapter 5. On the other hand, these cascaded integrators provide robustness, i.e., insensitivity to coupling or interaction effects. The transform of the response for the system in Figure 4-3 is

$$c(s) = \frac{KG_p(s)}{s + (D + sk_D(s))\,G_p(s)}\; r(s) \qquad (4\text{-}14)$$

We note that the integral error plus state feedback compensator does not add any zeros to the transfer function. Moreover, this compensator, in its various forms, provides complete pole assignment capability.

Comparison Procedure

The comparison of the control strategies for a multivariable system is based upon a plant transfer function given by Niederlinski (17):

$$[G(s)] = \frac{1}{(1 + .1s)\,(1 + .2s)^2} \begin{bmatrix} 1. & \dfrac{2.4}{1 + .5s} \\ \dfrac{.5}{1 + .1s} & 1. \end{bmatrix} \qquad (4\text{-}15)$$

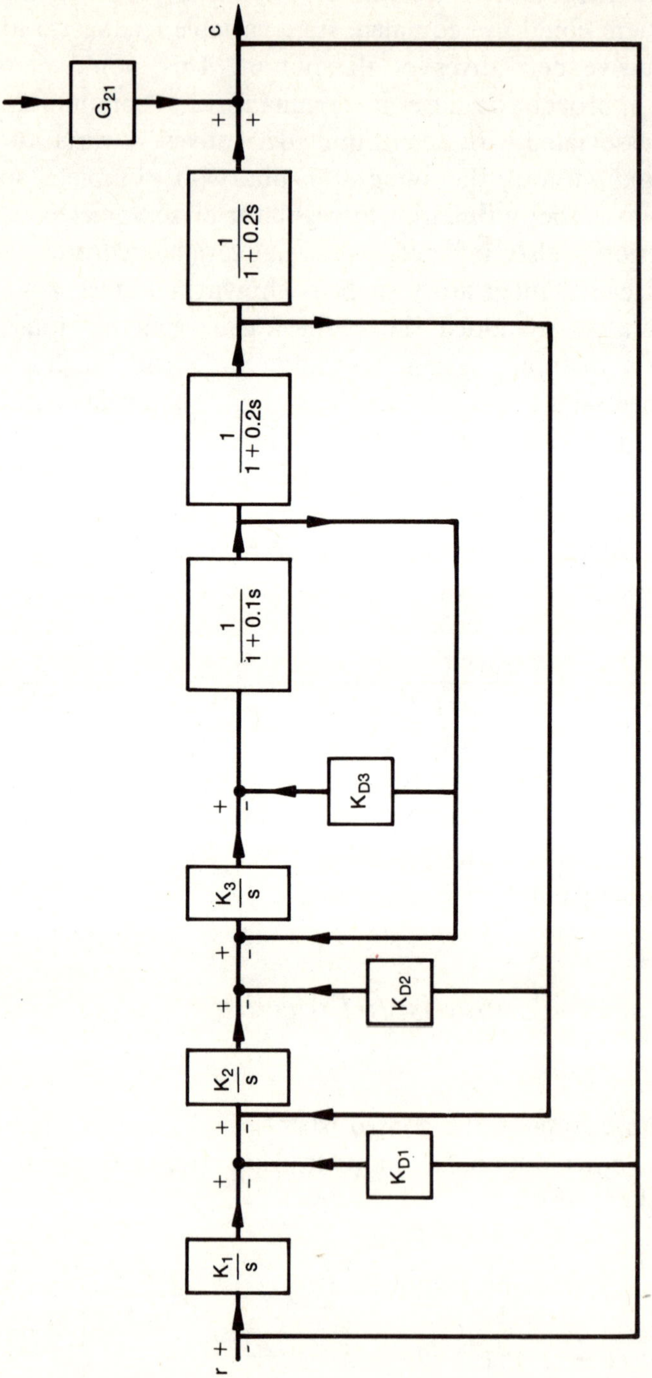

FIGURE 4-4
Three-pole plant with IESF (Integral error plus state feedback) control

This example was selected because the Ziegler-Nichols rules were used for selecting the proportional-integral-derivative feedback gains for a multivariable system (15).

Comparisons are made for the following:

1. Integral error plus state feedback control with three cascaded integrators and six repeated roots at $s = -25$
2. Proportional-integral-derivative control with Ziegler-Nichols gains
3. Proportional-integral-derivative control with integral absolute value of the error optimum gains

For cases 2 and 3, comparisons are made also for the system with a decoupling filter and without a decoupling filter. For the first case, decoupling filters are not used because the system performed with almost no interaction without the decoupling filter. Further comparisons are made to determine the effects of changing the plant time constants 25% while leaving unchanged the feedback gains and the decoupling gains. This is one way to take into account uncertainties about the plant parameters, modeling inaccuracies, and changes in materials properties due to temperature wear and aging. This provides an estimate of system robustness. The form of coupling between the two inputs and the two outputs did not permit a direct application of integral error plus state feedback compensation. It was decided, therefore, to consider the coupling "inputs" as disturbances to two single-input/single-output systems and then use integral error plus state feedback compensation with three cascaded integrators as shown in Figure 4-4. The cascaded integrators with complete state variable feedback allow for pole assignment. Response to a unit step input was simulated using CSMP (14). A step input at r_2 was found to be the more severe case and was used in the simulations.

Results

Table 4-1 contains common measurements of control performance for interacting systems and for decoupled systems with integral error plus state feedback or proportional-integral-derivative compensation. Time response curves for these cases are presented also.

TABLE 4-1
Performance Criteria for PID and IESF Controllers.

Control Type	Off-Design	Decoupling Filter	Gains K_C	T_I	T_D	Rise Time	Settling Time	Peak Overshoot %	IAE r_2/c_1	r_2/c_2
PID	N	N	6.4	0.364	0.091	0.24	1.36	47	0.211	0.351
Z-N	Y	N				0.28	4.25	65	0.675	0.773
	N	Y	6.4	0.364	0.091	0.22	0.76	51	0.0	0.303
	Y	Y				0.26	2.86	75	0.232	0.684
PID	N	N	6.4	2.0	0.11	0.28	0.56	7	0.314	0.210
Min	Y	N				0.32	0.94	16	0.317	0.288
IAE	N	Y	6.4	2.0	0.11	0.26	0.55	10	0.0	0.205
	Y	Y				0.30	1.01	24	0.043	0.299
IESF			$K_1=10.82$ $K_{D1}=1.6$							
	N	N				0.38	0.43	0	3.E-4	0.240
			$K_2=112.5$ $K_{D2}=5.77$							
	Y	N				0.38	0.40	4	3.5E-4	0.260
			$K_3=795.0$ $K_{D3}=13.0$							

Note: Off-design results are for time constant changes of 25%.

Proportional-Integral-Derivative Control with Ziegler-Nichols Gains

The gains K_c, T_I, and T_D were determined in Reference 17 for the interacting system. The r_1 and r_2 controllers use the same values. When the decoupling filter is added, two single-input/single-output systems result and the control gains are recalculated. It was found that the new values were only little different from the original ones, but they did not result in better response. We decided to use the original values (interacting control) for both cases. Time response for the interacting case, Figure 4-5, shows considerable interaction through c_1 and substantial overshoot for c_2. The addition of the decoupling filter completely decouples c_1 but still allows about the same overshoot for c_2 as shown in Figure 6. Time constant changes of 25% produced significant changes in the time response as shown in Figures 4-7 and 4-8. In particular the decoupling filter did little to reduce interaction effects. Certainly, the filter was designed for one plant and could not handle changes in the plant. This illustrates one of the difficulties associated with decoupling filters.

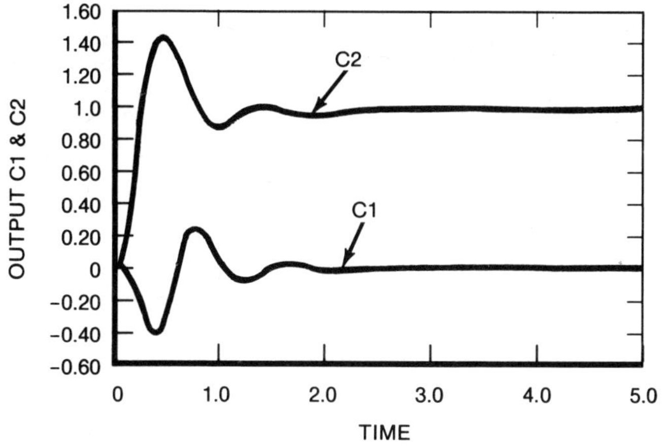

FIGURE 4-5
Response of interacting system with PID (Proportional-integral-derivative) control, Ziegler-Nichols gains. Unit step input at r_2.

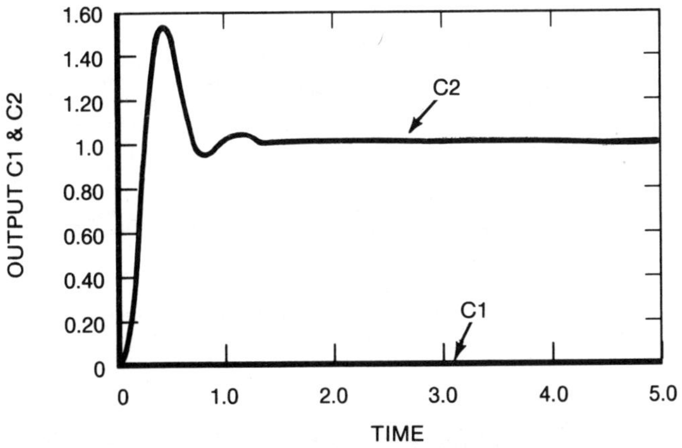

FIGURE 4-6
Response of decoupled system with PID (Proportional-integral-derivative) control, Ziegler-Nichols gains. Unit step input at r_2

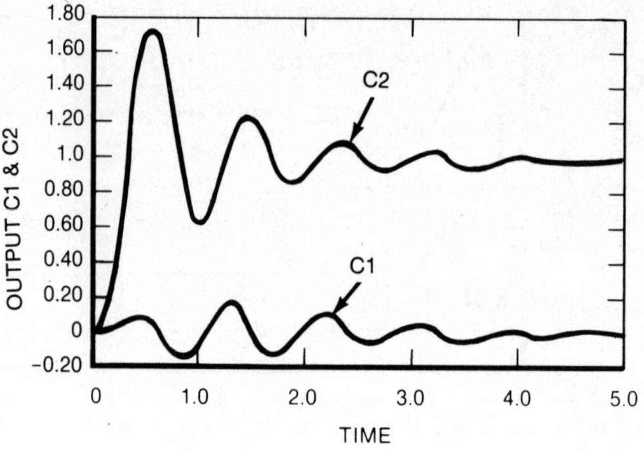

FIGURE 4-7
Response of interacting system with PID (Proportional-integral-derivative) control, Ziegler-Nichols gains. Plant time constants changed 25%. Unit setp input at r_2.

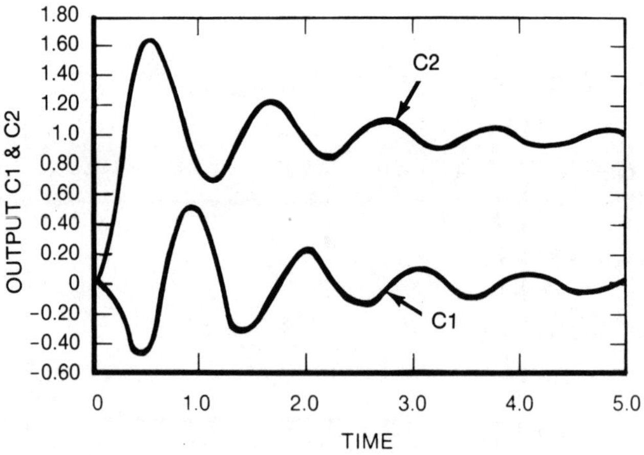

FIGURE 4-8
Response of decoupled system with PID control, Ziegler-Nichols gains. Plant time constants changed 25%. Unit step input at r_2.

Proportional-Integral-Derivative Control for Minimum Integrated Absolute Value of the Error

The integral of the absolute value of the error criterion is

$$\text{IAE} = \int_0^\infty |r-c|\, dt \tag{4-16}$$

Just as was found with the Ziegler-Nichols case, there was not much difference between optimum gain values of the interacting systems or the decoupled systems. Hence, the decoupled optimum values were used with the interacting case. It should be pointed out that if the integrated absolute value of the error were used for the disturbance r_1 to c_2 or r_2 to c_1, different optimum values would have resulted. Also, there are other integrals (11) such as integrated square of error or integral of time multiplied by absolute value of error. We selected the integral of the absolute value of the error criterion and used it for the proportional-integral-derivative controller. Also, we made IAE calculations for all runs to establish another comparative measure of performance.

Figures 4-9 and 4-10 show the time response without and with the decoupling filter, respectively. We note considerable improvement over the cases that used the Ziegler-Nichols gains. Similarly, the interaction and overshoot are changed less for the plant parameter changes as shown in Figures 4-11 and 4-12.

Integral Error Plus State Feedback Control

All state variables are considered to be available for feedback. They are the output of each first-order pole. The gains are selected by assuming two single-input/single-output systems where the original coupling is considered to be a disturbance. The six gains were selected to assign the six poles on the real axis at $s = -25$. Decoupling filters are not used. Time responses,

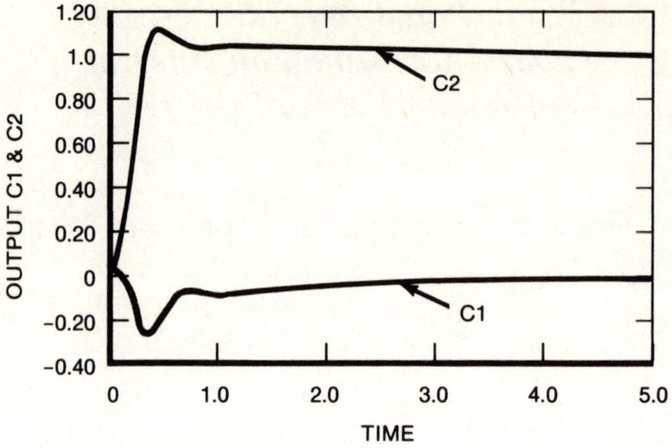

FIGURE 4-9
Response of interacting system with PID control, minimum IAE gains. Unit step input at r_2.

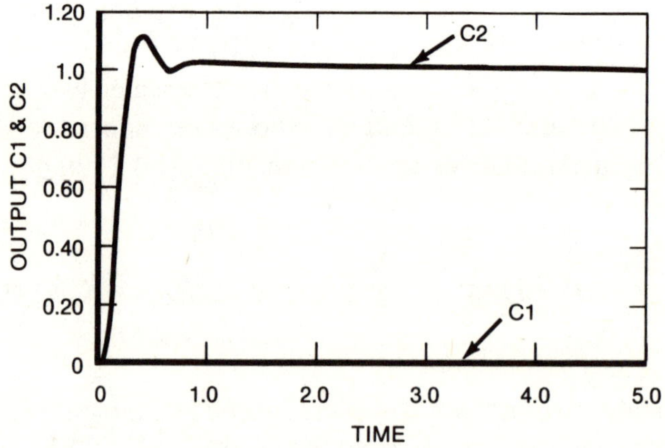

FIGURE 4-10
Response of decoupled system with PID control, minimum IAE gains. Unit step input at r_2.

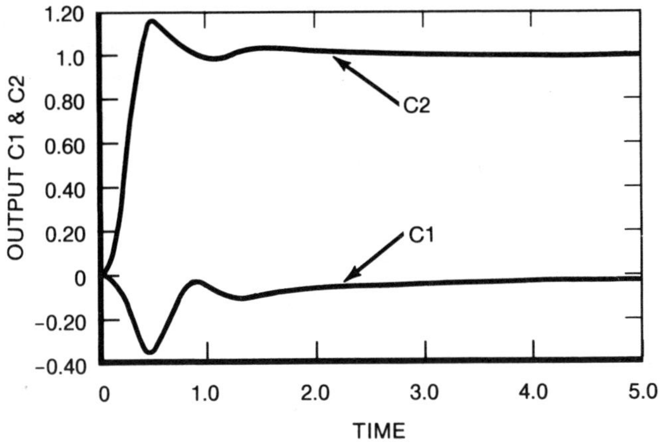

FIGURE 4-11
Response of interacting system with PID control, minimum IAE gain. Plant time constants changed 25%. Unit step input at r_2.

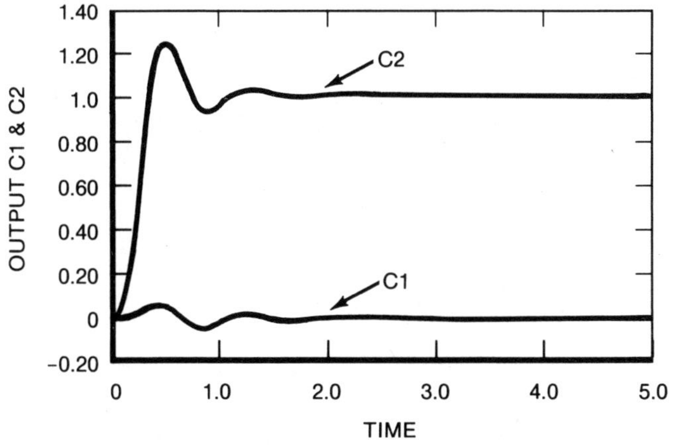

FIGURE 4-12
Response of decoupled system with PID control, minimum IAE gains. Plant time constants changed 25%. Unit step input at r_2.

Figures 4-13 and 4-14, show that interaction is virtually nonexistent, and off-design performance was little changed from the original case. We have achieved noninteracting response through the use of cascaded integrators and state variable feedback and without the use of decoupling filters.

Discussion

A control algorithm based upon state variable feedback and cascaded integrators has been shown to eliminate or substantially reduce interaction in multivariable systems with significant coupling. Decoupling filters are not required. Moreover, the system performs well even when the plant time constants are changed 25%. Numerical simulations show the superiority of this approach over the proportional-integral-derivative controller.

The algorithm permits complete freedom in pole assignment. For low-order systems this is a straightforward exercise, but as the order increases beyond eight or ten, the algebraic operations (e.g., determinant expansion) become burdensome. Much of this tedious activity can be eliminated, however, with one of the algebraic manipulation languages.

References

1. Anderson, B.D.O., and Moore, J.B., *Linear Optimal Control*, Prentice-Hall, Englewood Cliffs, N.J., 1971.
2. Athans, M., Dertouzos, M.L., Spann, R.N., and Mason, S.J., *Systems, Networks, and Computation: Multivariable Methods*, McGraw-Hill, New York, 1974.
3. Callier, F.M., and Desoer, C.A., *Multivariable Feedback Systems*, Springer-Verlag, New York, 1982.
4. Chen, C.T., *Introduction to Linear System Theory*, Holt, Rinehart, and Winston, New York, 1970.
5. Foss, A.M., *Multivariable Limit Control With Application to Gas Turbines*, Ph.D. Thesis, London University, 1981.
6. Fossard, A.J., *Multivariable System Control*, North Holland Publishing Company, Amsterdam, 1977.

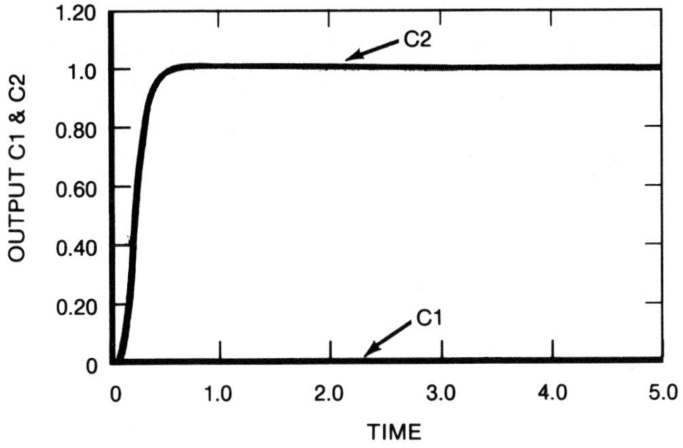

FIGURE 4-13
Response of coupled system with IESF control. Unit step input at r_2.

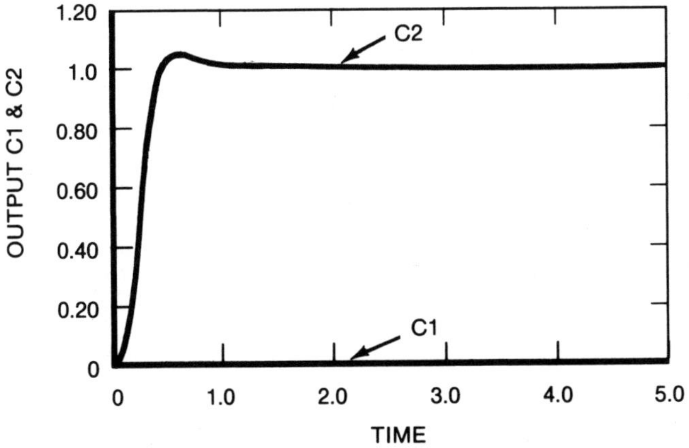

FIGURE 4-14
Response of coupled system with IESF control. Plant time constants changed 25%. Unit step input at r_2.

7. Hung, Y.S., and MacFarlane, A.G.J., *Multivariable Feedback, A Quasi-Classical Approach*, Springer-Verlag, Berlin, 1982.
8. Owens, D.H., *Multivariable and Optimal Systems*, Academic Press, London, 1981.
9. Patel, R.V., and Munro, N., *Multivariable System Theory and Design*, International Series on System and Control, Vol. 4, Pergamon Press, Oxford, 1982.
10. Safonov, M.G., *Stability and Robustness of Multivariable Feedback Systems*, MIT Press, Cambridge, Massachusetts, 1980.
11. Takahaski, Y., Rabins, M.J., and Auslander, D.M., *Control and Dynamic Systems*, Addison-Wesley, Reading, Massachusetts, 1970.
12. Phelan, R.M., *Automatic Control Systems*, Cornell University Press, Ithaca, N.Y., 1977.
13. Niederlinski, A., "Two-Variable Distillation Control: Decouple or Not Decouple," *AI Ch E Journal*, Vol. 17, No. 5, pp 1261–1263, 1971.
14. Continuous Systems Modeling Program (CSMP), Program No. 5734-X59, IBM.
15. Ziegler, J.G., and Nichols, N.B., "Optimum Settings for Automatic Controllers," *Transactions of the ASME*, Vol. 64, pp 759–768, 1942.
16. Phelan, R.M., *Subvariable Control*, Copyright 1981.
17. Niederlinski, A., "A Heuristic Approach to the Design of Linear Multivariable Interacting Control Systems," *Automatica*, Vol. 7, pp 691–701, 1971.
18. Maday, C.J., "Synthesis and Design of Feedback Control Systems for Time Response I," Proceedings of the 1984 American Control Conference, San Diego, CA, June 6–8, 1984.

CHAPTER V

Pole Assignment in Hybrid Analog/Discrete Systems

Introduction

Efficient and general procedures are developed to analyze, synthesize, and design multivariable digital feedback control systems for time response. The objective is to develop an algorithm to achieve exact pole assignment in large scale differential/difference systems where a time-invariant, time-continuous plant is controlled by a digital computer. The plant is represented in discrete form with transition matrices over the sampling interval. Error compensation is a combination of state feedback and Euler integration. This allows the entire system to be represented exactly in the z-plane. Pole assignment is made along the real axis $0 \leq z < 1$. An algebraic manipulation language is used to expand large determinants and construct computer programs for the calculation of system gains. Computer simulations, made to evaluate the effectiveness of the algorithms, demonstrate very robust system performance with respect to external disturbances and changes in the plant parameters and the sampling interval.

Algorithms are available for pole assignment in systems with state feedback. For very large systems the coefficient matrix for the unknown gains may be ill-conditioned. An alternative scheme uses the algebraic expansion of the system determinant for the gain calculations, a tedious job with many opportunities for error. Most of the work, and any possible errors, can be eliminated by using an algebraic manipulation language. The system determinant is expanded algebraically and ordered in powers of z by the computer. Then, a FORTRAN program can be generated by the algebraic manipulation language to obtain the required feedback gains and any integrator gains. This is an automated procedure that does not require manual transcription of intermediate results.

Objective

The overall objective is to develop a general and useful procedure for pole assignment in multivariable digital feedback control systems. The emphasis on digital control stems from both practical and theoretical considerations. High performance control algorithms may require gains up to 1000. Analog components are typically limited to 10 V peak voltage, so the noise problem associated with the control signal at high amplification will limit accuracy. Also, digital controllers are easily modified during the developmental period and allow alternate control schemes to be implemented in software rather than in the hardware.

State of the Art

The theory and design of multivariable control systems continues to be an area of active research that has produced important results (1 through 11). The development of the theory and design of these systems has proceeded along several lines that are parallel and somewhat related. The dominant feature common to these approaches is the use of state feedback to achieve some design objective. State feedback, optimal for the linear quadratic regulator (LQR) or servomechanism, has been useful also in the design of control systems by pole assignment. In large systems, however, pole assignment algorithms may have trouble with ill-conditioned systems of algebraic equations in the unknown gains.

It has been pointed out (9) that one of the aspects of linear multivariable system analysis and design that requires further investigation is the development of efficient computational algorithms. This is important because the success of theoretical studies and design methods depends upon their numerical implementation. The work described here addresses the pole assignment problems and the simulation problems associated with the larger multivariable systems.

Problem Statement

This work builds on the state of the art described above and concentrates on the computational aspects of multivariable feedback control system synthesis, analysis, and design. Emphasis is placed on the design of sampled data systems in the z-domain through pole assignment.

Consider a time-continuous, time-invariant plant controlled by a digital computer that implements a control algorithm. State feedback and cascaded stages of integral compensation characterize the control. The piecewise continuous output of the direct digital controller (DDC) drives the time-continuous plant. Changes in the output of the controller occur at each sampling interval. The vector/matrix block diagram in Figure 5-1 illustrates the general features of the overall system. The system dynamics are characterized simultaneously by difference equations that describe the control algorithm and by differential equations that describe the time-continuous plant. In the design of a feedback control system, the desired time response or frequency response is obtained by selecting appropriate values for the state feedback gains, the integral gains, and possible derivative gains. The design tools common to analog systems include the root-locus plot, the Bode diagram, the Nyquist

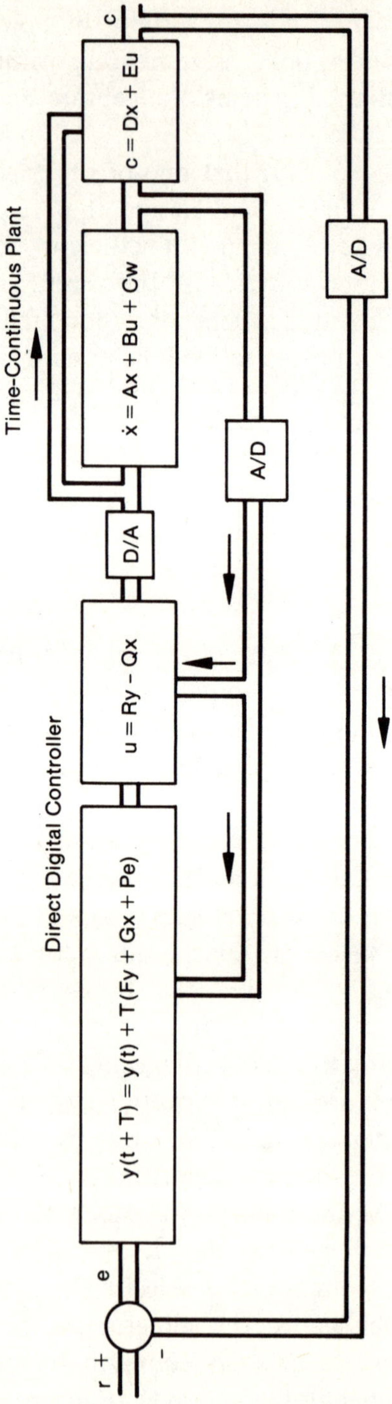

FIGURE 5-1
Vector-matrix block diagram of continuous plant controlled by a discrete controller.

diagram, and pole assignment. Completely digital control system design tools are similar except that they are used in the z-plane rather than the s-plane. The mixed character of the system under consideration, however, complicates attempts to apply these design tools successfully. Moreover, in multivariable systems having several unspecified gains, implementation of most of these tools either is cumbersome and relatively ineffective or requires extensive software and interactive graphics capability. For example, the root-locus plot is effective for one unknown gain, cumbersome for two unknown gains (a surface plot is required) and generally interactable for three or more gains (hypersurface plots would be required). For multivariable systems with state feedback, the linear quadratic regulator (or servomechanism) is a valuable approach that gives excellent results. The cost of determining the $n(n+1)/2$ components of the matrix Riccati differential equation for an nth-order system must be weighed against the benefits realized from this approach. Pole assignment can be a useful design tool for relatively low-order systems; cumbersome algebraic manipulations or ill-conditioned systems characterize the orders. Much of the algebraic work can be facilitated, however, by the use of symbolic computer languages. These remarks apply to continuous systems and to discrete systems equally.

BACKGROUND

The difficulties associated with differential-difference systems follow from time delays inherent in digital systems. For linear systems, the Laplace transform produces transfer functions in the s-domain for continuous systems and transfer functions in the z-domain for discrete systems. Attempts to use frequency domain techniques or pole assignment methods have encountered difficulty stemming from the simultaneous s and z representations in the same dynamic system.

A number of approximate methods have been suggested to represent the entire system in the z-plane (12, 13). The z-plane is appropriate because it handles the pulse character of digital systems much better than does the s-plane. For a mixed system, each transfer function is obtained either in terms of s or z as dictated by the continuous character or by the discrete character of each component. Then, all terms in s are converted to an approximate z representation. Since there are an infinite number of z representations for s^n, a decision must be made about how approximately the continuous components can be represented. Available approximations include a pulse train, piecewise constant segments, trapezoidal segments, or parabolic segments. The pulse train and the piecewise constant representations are usually inadequate. The trapezoidal form is generally used because the more general parabolic forms involve burdensome computation. The sensitivity of these approximations to the sampling frequency remains a consideration, however.

Although the use of a direct digital controller strongly suggests analysis and design in the z-plane, another approximation considers the system as entirely continuous for design purposes. Feedback gains and feedforward gains are determined upon this assumption. For sampling intervals that are short relative to system time constants, this approach can give good results. It will be shown that it can also produce instability.

Solution Technique

Effective design techniques depend upon the representation of each component of the system in the z-plane. The direct digital controller can be represented exactly because it is a discrete device. Where integral compensation is used, the actual integration may be approximated by the discrete representation if the approximate process is exactly representable. This feature is used to good advantage in the implementation of pole assignment techniques. Moreover, it will be shown that it is better to describe an approximate algorithm exactly than it is to describe an exact algorithm approximately.

Since we consider only a time-invariant and continuous plant,

$$\dot{\underset{\sim}{x}} = \underset{\sim}{A}\,\underset{\sim}{x} + \underset{\sim}{B}\,\underset{\sim}{u} + \underset{\sim}{C}\,\underset{\sim}{w} \qquad (5\text{-}1)$$

where $\underset{\sim}{w}$ is any external disturbance.

The transition and input matrices $\underset{\sim}{\Phi}$ and $\underset{\sim}{\psi}$ are given by

$$\underset{\sim}{\Phi}(t, t_o) = L^{-1}[(\underset{\sim}{I} - \underset{\sim}{A})^{-1}] \qquad (5\text{-}2)$$

$$\underset{\sim}{\psi}(t, t_o) = \int_{t_o}^{t} \underset{\sim}{\Phi}(t - \tau)\underset{\sim}{B}\underset{\sim}{u}(\tau)d\tau \tag{5-3}$$

L is the symbol for the Laplace transform. The solution to Equation 5-1 is

$$\underset{\sim}{x}(t) = \underset{\sim}{\Phi}(t, t_o)\underset{\sim}{x}(t_o) + \underset{\sim}{\psi}(t, t_o) \tag{5-4}$$

where the interval $[t, t_o]$ is arbitrary. For sampled data systems, $[t, t_o]$ can be sampling interval T, over which the control $u(t)$ is constant. That is, the control is piecewise constant. We take advantage of this in Equation 5-3,

$$\begin{aligned} \underset{\sim}{\psi}(t, t_o) &= \int_{t_o}^{t} \underset{\sim}{\Psi}(t - \tau)\underset{\sim}{B}d\tau \; \underset{\sim}{u}(t_o) \\ &= \underset{\sim}{\theta}(t, t_o)\underset{\sim}{u}(t_o) \end{aligned} \tag{5-5}$$

where $\underset{\sim}{\theta} = \int_{t_o}^{t} \underset{\sim}{\Phi}(t - \tau)\underset{\sim}{B}d\tau$

The solution to Equation 5-1 becomes

$$\underset{\sim}{x}(t) = \underset{\sim}{\Phi}(t, t_o)\underset{\sim}{x}(t_o) + \underset{\sim}{\theta}(T)\underset{\sim}{u}(t_o) \tag{5-6}$$

For the sampled-data systems, it is

$$\underset{\sim}{x}(t + T) = \underset{\sim}{\Phi}(T)\underset{\sim}{x}(t) + \underset{\sim}{\theta}(T)\underset{\sim}{u}(t) \tag{5-7}$$

The system output is

$$\underset{\sim}{c} + \underset{\sim}{D}\underset{\sim}{x} + \underset{\sim}{E}\underset{\sim}{u} \tag{5-8}$$

The direct digital controller is described by the difference equation

$$\underset{\sim}{y}(t + T) = \underset{\sim}{y}(t) + T(\underset{\sim}{E}\underset{\sim}{y} + \underset{\sim}{G}\underset{\sim}{x} - \underset{\sim}{P}\underset{\sim}{r}) \tag{5-9}$$

The control is a linear combination of the state and the controller variables,

$$\underset{\sim}{u} = \underset{\sim}{R}\underset{\sim}{y} - \underset{\sim}{Q}\underset{\sim}{x} \tag{5-10}$$

The system described by Equations 5-7 through 5-10 is z-transformed

$$\begin{aligned} (z\underset{\sim}{I} - \underset{\sim}{\Phi})\underset{\sim}{x} &= \underset{\sim}{\theta}\,\underset{\sim}{u} \\ \underset{\sim}{c} &= \underset{\sim}{D}\underset{\sim}{x} + \underset{\sim}{E}\underset{\sim}{u} \\ [(z-1)\underset{\sim}{I} - T\underset{\sim}{F}]\underset{\sim}{y} &= T(\underset{\sim}{G}\underset{\sim}{x} + \underset{\sim}{P}\underset{\sim}{r}] \\ u &= \underset{\sim}{R}\underset{\sim}{y} - \underset{\sim}{Q}\underset{\sim}{x} \end{aligned} \tag{5-11}$$

and is represented in the z-plane only. The matrices $\underset{\sim}{F}$, $\underset{\sim}{G}$, $\underset{\sim}{R}$, and $\underset{\sim}{Q}$ contain feedback gains and dynamic element (integrators, for example) gains, which are computed for pole assignment.

The matrix representation of the solution to Equation 5-1 is exact although the state $\underset{\sim}{x}(t)$ is computed at discrete points in time. The effect is that of viewing the system with a flashing stroboscopic light. The plant output is continuous in time, but the output is viewed or sampled at specified time intervals. It is this feature of the transition matrices that allows the time-continuous plant to be represented exactly with difference equations.

EXAMPLE

The control of a spring-mass system with viscous damping is considered. Position and velocity are accessible. The DDC contains two cascaded stages of approximate integration that use the Euler method of computation. The state equations, in discrete form, for the block diagram in Figure 5-2 are:

$$y_1(t+T) = y_1(t) + K_1 T[r(t) - x_2(t)] \tag{5-12}$$

$$y_2(t+T) = y_2(t) - K_2 T[y_1(t) - K_{D1}x_2(t) - x_1(t)] \tag{5-13}$$

$$x_1(t+T) = \Phi_{11}x_1(t) + \Phi_{12}x_2(t) + \theta_1[y_2(t) - K_{D2}x_1(t)] \tag{5-14}$$

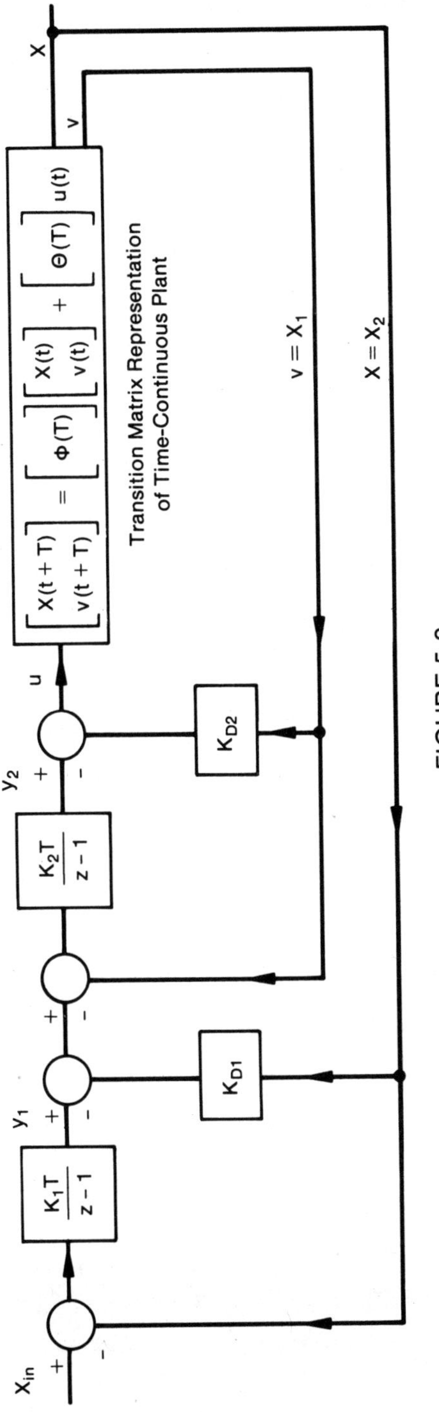

FIGURE 5-2
Block diagram of complete system—spring/mass plant controlled by a digital controller. Integral compensation is by Euler integration.

$$x_2(t+T) = \Phi_{21}x_1(t) + \Phi_{22}x_2(t) + \theta_2[y_2(t) - K_{D2}x_1(t)] \tag{5-15}$$

The z-transform of this system of equations is:

$$y_1(z-1) + K_1Tx_2 = K_1Tr \tag{5-16}$$

$$y_2(z-1) + K_2Ty_1 - K_2Tx_1 - K_2TK_{D1}x_2 = 0 \tag{5-17}$$

$$x_1(z - \Phi_{11} + K_{D2}\theta_1) - \Phi_{12}x_2 - \theta_1y_2 = 0 \tag{5-18}$$

$$x_2(z - \Phi_{22}) - x_1(\Phi_{21} - K_{D2}\theta_2) - \theta_2y_2 = 0 \tag{5-19}$$

and the determinantal or characteristic equation, obtained with the aid of REDUCE™, is:

$$\begin{aligned} &z^4+z^3(K_{D2}\theta_1-\Phi_{22}-\Phi_{11}-2)+z^2[K_2K_{D1}\theta_2T+K_2\theta_1T-K_{D2}(\theta_{22}-\Phi_{12}\theta_2+2\theta_1) \\ &\quad +\Phi_{22}\Phi_{11}+2\Phi_{22}-\Phi_{21}\Phi_{12}-2\Phi_{11}+1] + z[K_1K_2T^2\theta_2+K_{D1}K_2T(\Phi_{21}\theta_1-\theta_2\Phi_{11} \\ &\quad -\theta_2)-K_2T(\Phi_{22}\theta_1-\theta_2\Phi_{12}+\theta_1)+K_{D2}(2\Phi_{22}\theta_1-2\theta_2\Phi_{12}+\theta_1)-2\Phi_{22}\Phi_{11} \\ &\quad -\Phi_{22}+2\Phi_{21}\Phi_{12}-\Phi_{11}]+K_1K_2T^2(\theta_1\Phi_{21}-\theta_2\Phi_{11})+K_{D1}K_2T(\theta_2\Phi_{11}-\theta_1\Phi_{21}) \\ &\quad +K_2T(\theta_1\Phi_{22}-\theta_2\Phi_{12})+K_{D2}(\theta_2\Phi_{12}-\theta_1\Phi_{22})+\Phi_{11}\Phi_{22}-\Phi_{12}\Phi_{21} = 0 \end{aligned} \tag{5-20}$$

This equation is shown here to emphasize the amount of algebra required for this relatively low-order system. Symbolic languages become attractive for moderately high-order systems.

EVALUATION OF THE TRANSITION AND INPUT MATRICES

It is evident from Equation 5-2 that for large systems a substantial amount of computation is required to evaluate the transition matrix $\underset{\sim}{\Phi}\,(t,t_o)$. The matrix inverse of $(s\underset{\sim}{I}-\underset{\sim}{A})$ is required so that the inverse Laplace transform can be determined. The evaluation of $\underset{\sim}{\theta}$ from Equation 5-5 also requires much computational effort. An alternate approach makes use of the definition of the transition and input matrices, which permits their direct numerical evaluation. From Equation 5-7 we note that $\underset{\sim}{\Phi}$ and $\underset{\sim}{\theta}$ are time-invariant and depend only upon the magnitude of the sampling interval. It is useful to expand

Equation 5-7 where $\underset{\sim}{x}$ is a two-component vector and u is a scalar. Then,

$$x_1(t+T) = \Phi_{11}(T)x_1(t) + \Phi_{12}(T)x_2(t) + \theta_1(T)u(t) \qquad (5\text{-}21)$$

$$x_2(t+T) = \Phi_{21}(T)x_1(t) + \Phi_{22}(T)x_2(t) + \theta_2(T)u(t) \qquad (5\text{-}22)$$

The Φ_{ij} and θ_k can be regarded as influence coefficients. For example, if we set $x_1(t) = 1$, $x_2(t) = u(t) = 0$ and integrate Equation 5-1 numerically from t to $t + T$, we would obtain

$$x_1(t+T) = \Phi_{11}(T)\cdot 1 \qquad (5\text{-}23)$$

$$x_2(t+T) = \Phi_{21}(T)\cdot 1 \qquad (5\text{-}24)$$

where it is convenient to select $t = 0$. Thus $\Phi_{11}\,(T)$ is the value of $x_1(T)$ for $x_2(0) = u = 0$, $x_1(0) = 1$, and Φ_{21} is the value of $x_2(T)$ for $x_2(0) = u = 0$, $x_1(0) = 1$. The process is repeated for $x_2(0) = 1$, $x_1(0) = u = 0$ so that we obtain Φ_{12} and θ_{22}. Finally, we set $u = 1$, $x_1(0) = x_2(0) = 0$ to obtain θ_1 and θ_2. Three separate numerical integrations of the system equations are required to obtain the elements of $\underset{\sim}{\Phi}$ and $\underset{\sim}{\theta}$ for a given sampling interval T. The direct evaluation of these elements makes this an attractive alternative for finding $\underset{\sim}{\Phi}$ and $\underset{\sim}{\theta}$. Pole assignment follows from Equation 5-20.

POLE ASSIGNMENT

Four coefficients of z^i, $i = 0, 1, 2, 3$, appear in Equation 5-20. Each coefficient is set equal to a numerical value consistent with the desired pole placement. For example, very fast deadbeat response to a step input is obtained by placing the poles at $z = 0$. Then each coefficient is zero and the four unknown gains K_1, K_2, K_{D1}, and K_{D2} can be determined from the four resulting equations. Although these gains appear nonlinearly, the system is linear in K_{D2}, K_2K_{D1}, K_2K_1, and K_2. Moreover, the system can be solved sequentially for K_{D2}, K_2K_{D1}, K_1K_2, and K_2. K_1 and K_{D1} follow directly. Any problems with ill-conditioned matrices are avoided. When slower deadbeat response is acceptable, the poles are placed on the real axis such that $0 < z < 1$. There is an approximate mapping between the s-plane and the z-plane for this case. A pole on the negative real axis at $\underline{a}$ in the s-plane maps onto the positive real axis at e^{-aT} in the z-plane where T is the sampling interval. This is an approximate mapping because system response depends upon the sampling interval T as well as the value of $\underline{a}$. For a given value of T, however, the mapping behaves as expected, i.e., as $\underline{a}$ increases the system becomes more responsive. For a fixed value of $\underline{a}$, however, the response is delayed as the sampling interval is increased, i.e., the response is shifted an amount proportional to the difference in the two sampling intervals.

Simulation Results

The spring-mass system has a normalized undamped natural frequency of unity. The damping ratio is specified interactively. The response is obtained for inputs of a unit step, a unit ramp, and a harmonic forcing function. Sampling intervals are specified interactively. Simulation results are shown in Figures 5-3 through 5-47. For all cases, pole location is on the positive real axis, $0 \leq z < 1$. For sufficiently small sampling intervals, this gives deadbeat response to a step input. Note that a normalized force of 39.48 causes a steady state displacement of unity.

A number of general conclusions can be drawn from these simulations. They include the following:

1. Sampling rates corresponding to the Nyquist frequency (twice the undamped frequency of the spring-mass system) must be avoided unless damping is present.
2. Sampling rates between the Nyquist frequency and twice the Nyquist frequency result in unusual behavior. For example, in Figures 5-26 and 5-27, the force is applied initially in the "wrong" direction. The reason for this is not known at the present.
3. Sampling rates at the undamped frequency of the spring-mass will work only if there is damping in the system.
4. Excellent frequency response can be expected if the poles are placed at $z = 0$ and the sampling frequency is greater than 5 times the forcing frequency.

5. Perhaps the most important conclusion is that it is possible to obtain excellent performance in digital systems with sampling frequencies that are as low as three times the open-loop frequency, or five times the forcing frequency. This is accomplished by designing entirely in discrete domain. Figure 5-3 illustrates what happens when the system is designed as an analog system but is operated in a discrete mode: instability results. Figure 5-4 shows the response for design in the discrete domain. Response is deadbeat as required.

Response to a step input is shown in Figures 5-5 through 5-16 for a number of sampling frequencies and pole locations. For poles assigned to $z = 0$, the mass moves the required amount in four sampling intervals. Nothing happens in the first two intervals because each of the integrators in the controller produce a single delay. This emphasized that integration, even in analog systems, cause delays.

Response to a unit ramp is shown in Figures 5-17 through 5-19. In general, the response is delayed about 2.5 sampling intervals. At sampling frequencies less than twice the Nyquist frequency, performance degrades.

Response to harmonic inputs is shown in Figures 5-20 through 5-24. In the fourth conclusion above it was noted that excellent response can be obtained with sampling intervals greater than 5 times the forcing frequency.

Response to a step disturbance is shown in Figures 5-25 through 5-31. Note that the steady-state error in all cases is zero. The force behavior for $T = 0.25$ (twice the Nyquist frequency) is interesting as a bang-bang-bang-bang.

Response to a harmonic disturbance is portrayed in Figures 5-32 through 5-35. In general, for sampling frequencies about ten times the disturbance frequency, the response is only about 2% of full scale.

Robustness properties are explored in Figures 5-36 through 5-46. In each case the system model was not as originally specified. For substantial changes in the undamped natural frequency, the damping ratio, and the sampling interval, stability is preserved except in Figure 5-39. Even here, if the poles were located at $z = 0.5$, performance would be satisfactory.

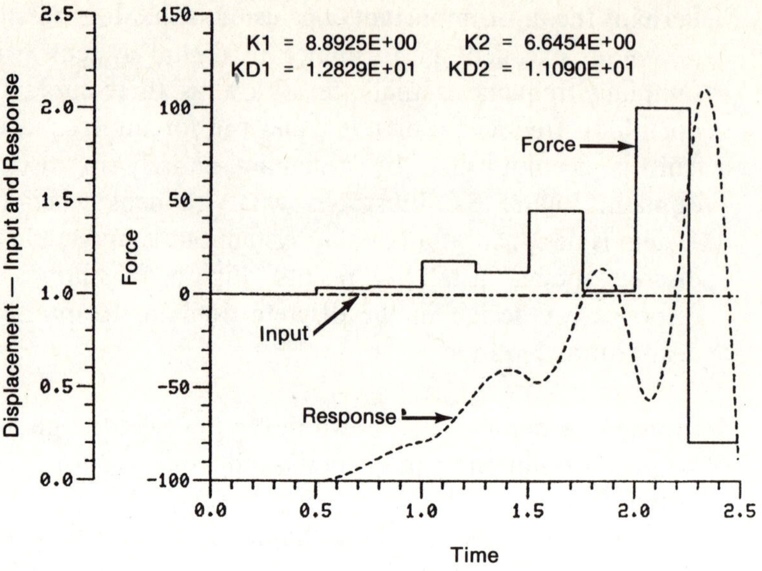

FIGURE 5-3
Response and actuator force for step input. ω_n = 1.0, ζ = 0, T = 0.25, repeated poles at z = 0.5. Gains calculated on the basis of continuous approximation.

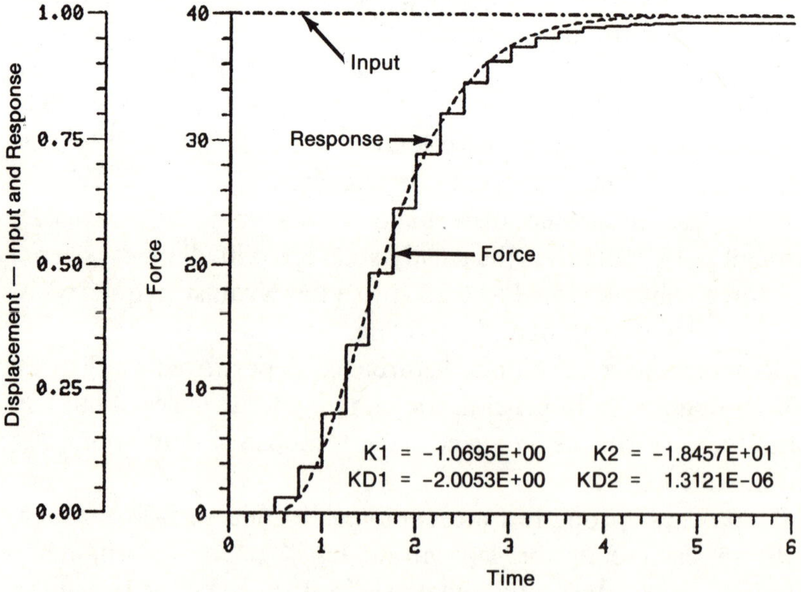

FIGURE 5-4
Response and actuator force for step input. ω_n = 1.0, ζ = 0, T = 0.25, repeated poles at z = 0.5. Compare with continuous approximation in Figure 5-3.

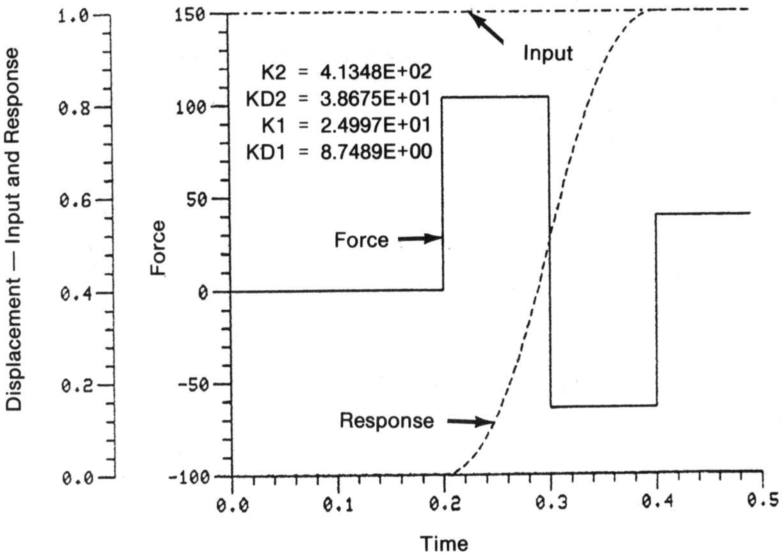

FIGURE 5-5
Response and actuator force for step input. ω_n = 1.0, ζ = 0, T = 0.1, repeated poles at z = 0.

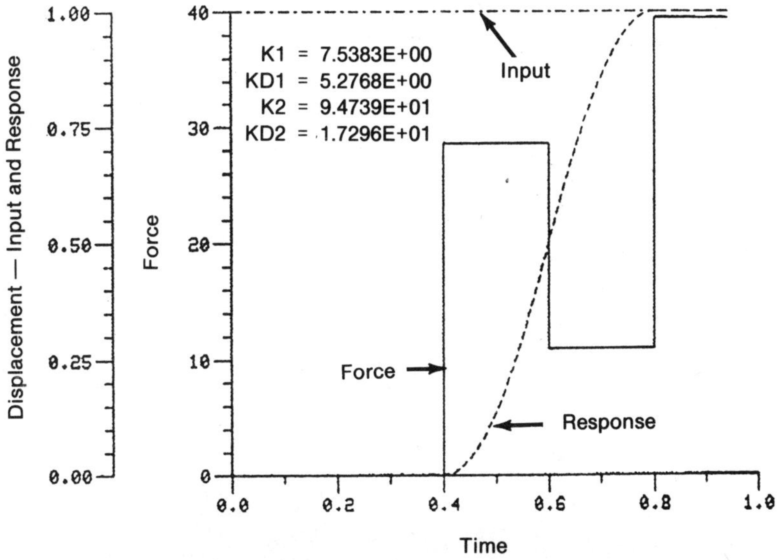

FIGURE 5-6
Response and actuator force for step input. ω_n = 1.0, ζ = 0, T = 0.2, repeated poles at z = 0.

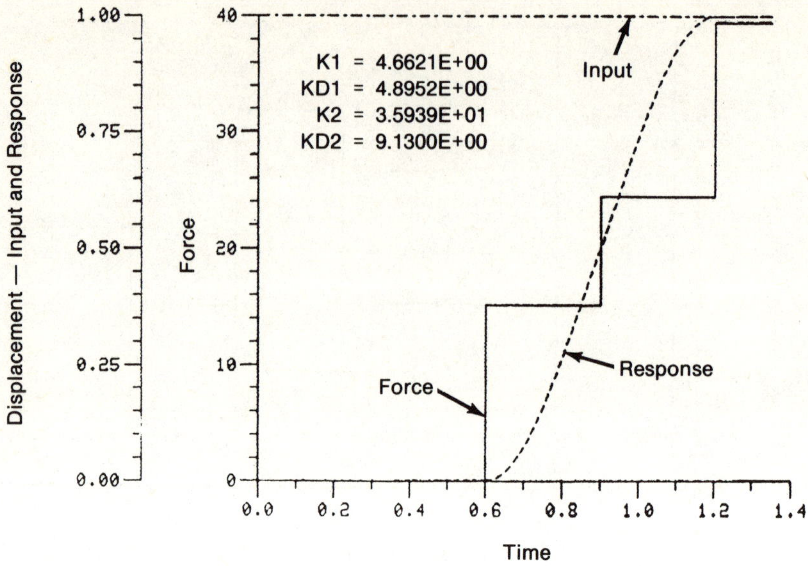

FIGURE 5-7
Response and actuator force for step input. ω_n = 1.0, ζ = 0, T = 0.3, repeated poles at z = 0.

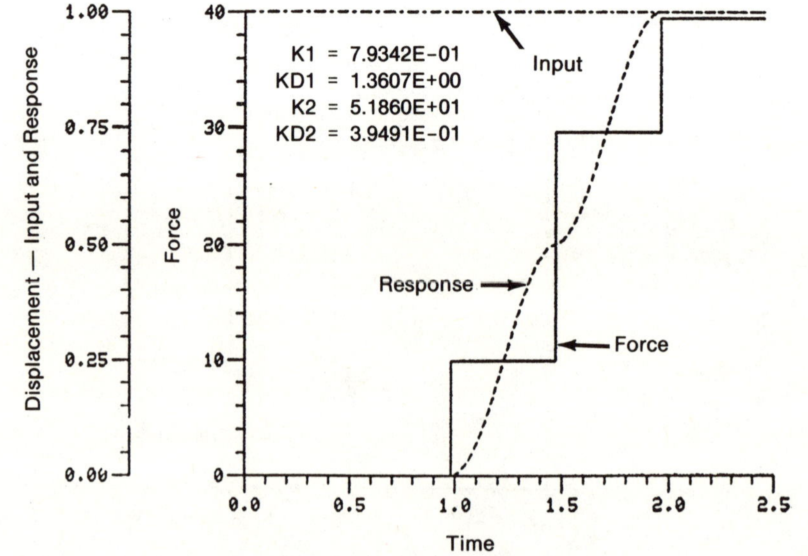

FIGURE 5-8
Response and actuator force for step input. ω_n = 1.0, ζ = 0, T = 0.49, repeated poles at z = 0.

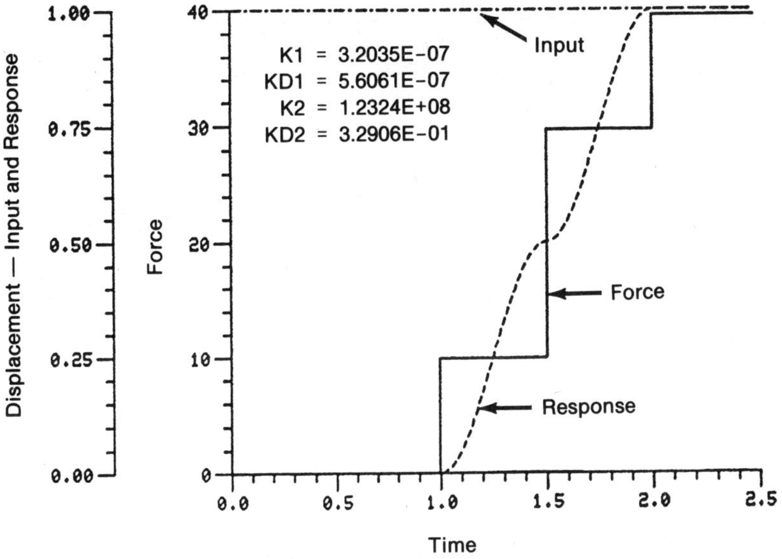

FIGURE 5-9
Response and actuator force for step input. $\omega_n = 1.0$, $\zeta = 0$, T = 0.5 (corresponds to Nyquist frequency), repeated poles at z = 0.

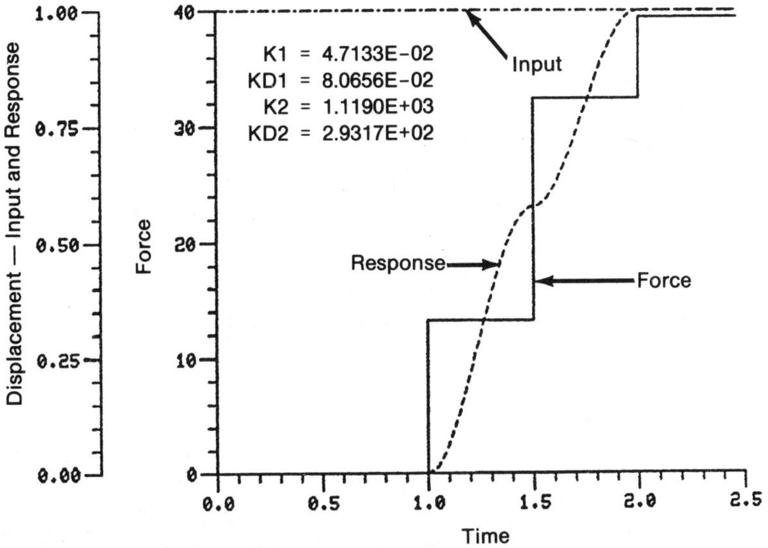

FIGURE 5-10
Response and actuator force for step input. $\omega_n = 1.0$, $\zeta = 0.1$, T = 0.5 (corresponds to Nyquist frequency), repeated poles at z = 0.

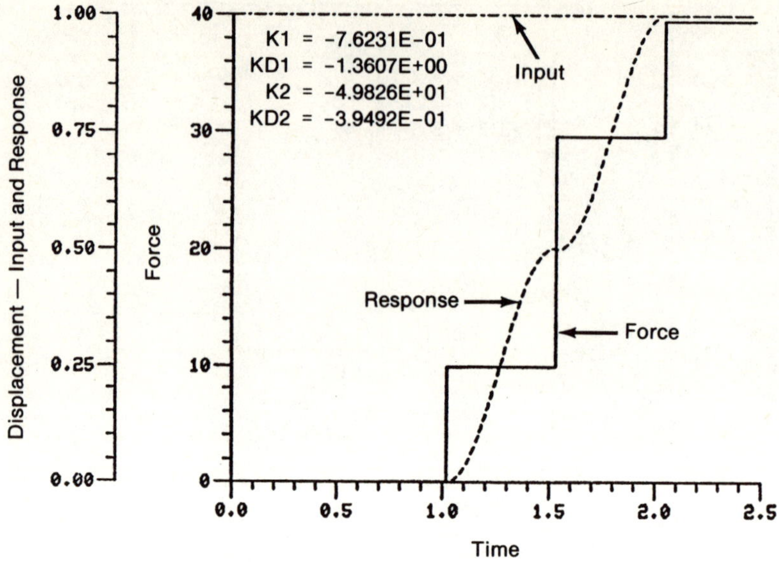

FIGURE 5-11
Response and actuator force for step input. ω_n = 1.0, ζ = 0, T = 0.51, repeated poles at z = 0.

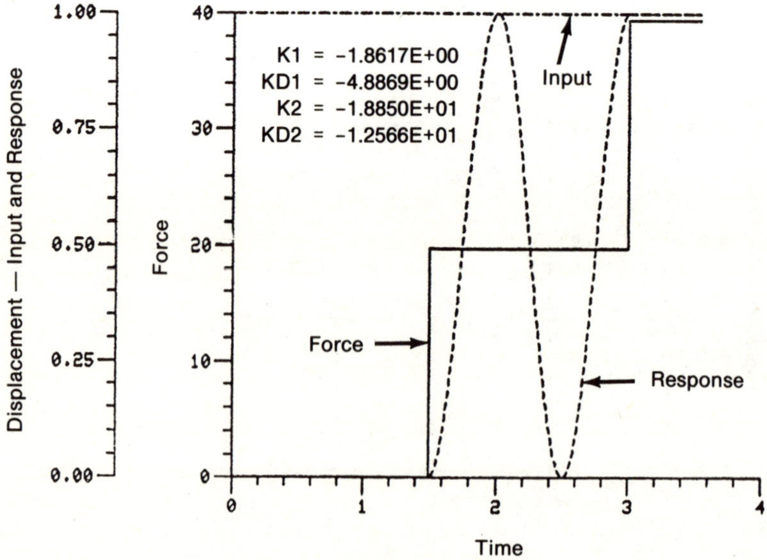

FIGURE 5-12
Response and actuator force for step input. ω_n = 1.0, ζ = 0, T = 0.75, repeated poles at z = 0.

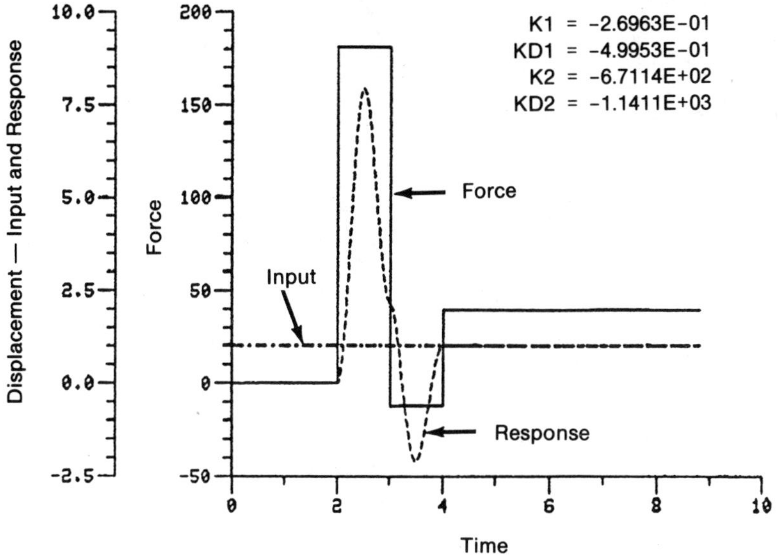

FIGURE 5-13
Response and actuator force for step input. $\omega_n = 1.0$, $\zeta = 0.1$, T = 1.0, repeated poles at z = 0.

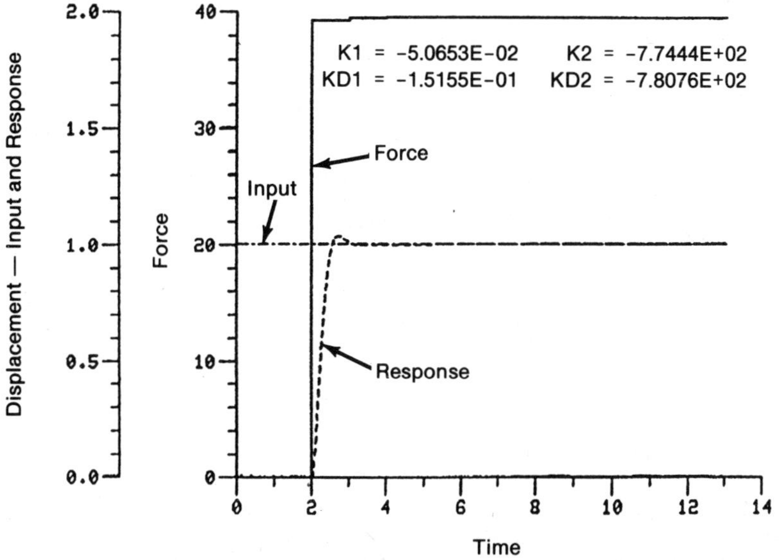

FIGURE 5-14
Response and actuator force for step input. $\omega_n = 1.0$, $\zeta = 0.707$, T = 1.0, repeated poles at z = 0.

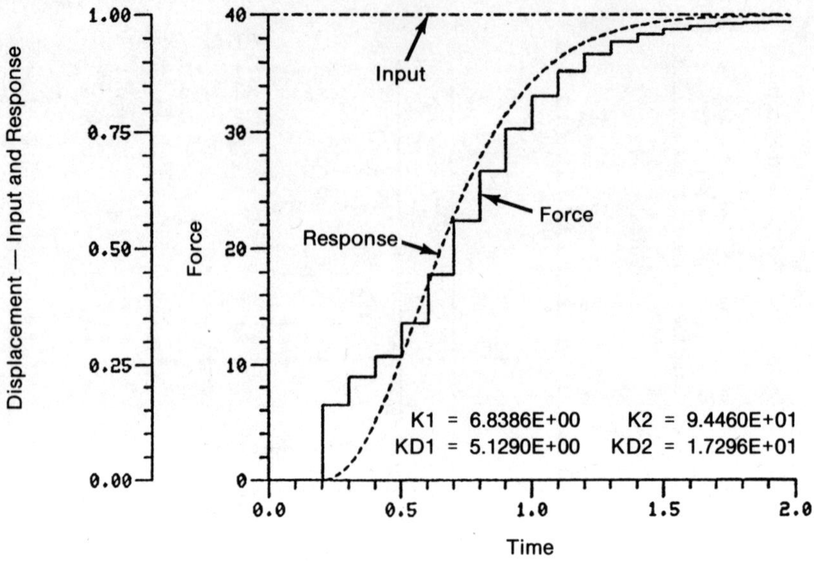

FIGURE 5-15
Response and actuator force for step input. ω_n = 1.0, ζ = 0, T = 0.1, repeated poles at z = 0.5.

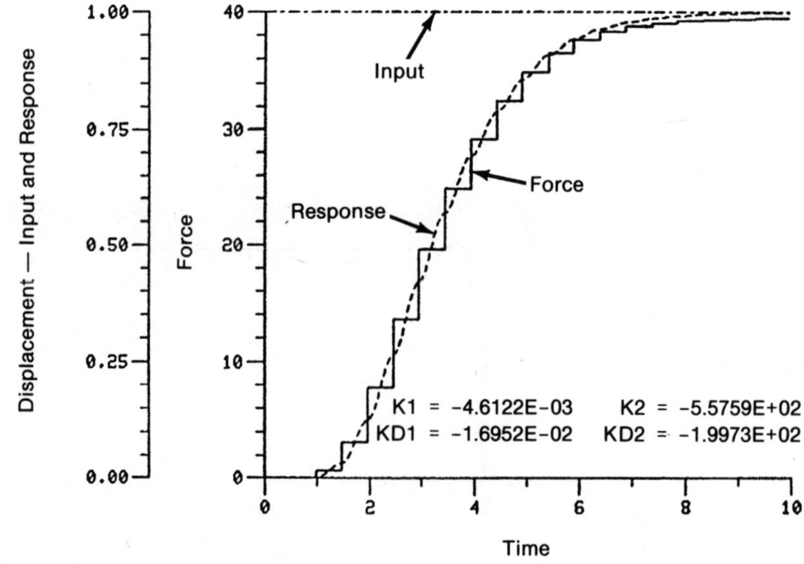

FIGURE 5-16
Response and actuator force for step input. ω_n = 1.0, ζ = 0, T = 0.49, repeated poles at z = 0.5.

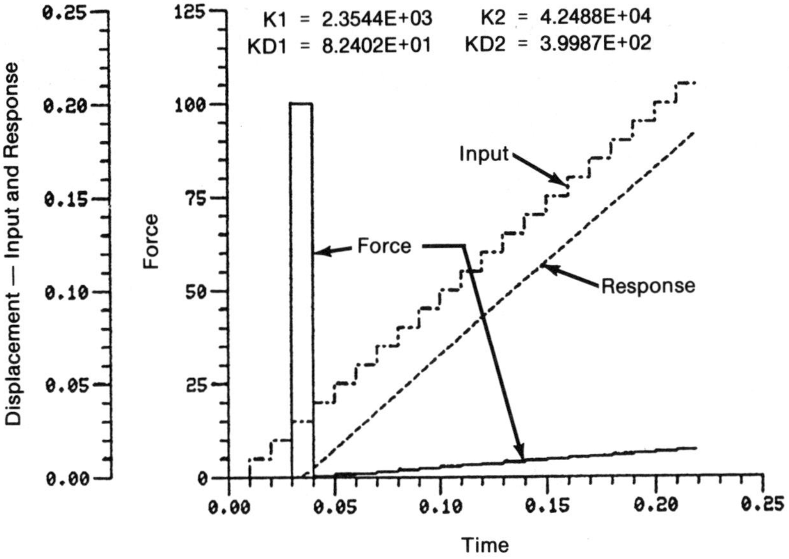

FIGURE 5-17
Response and actuator force for unit ramp input. $\omega_n = 1.0$, $\zeta = 0$, T = 0.01, repeated poles at z = 0.

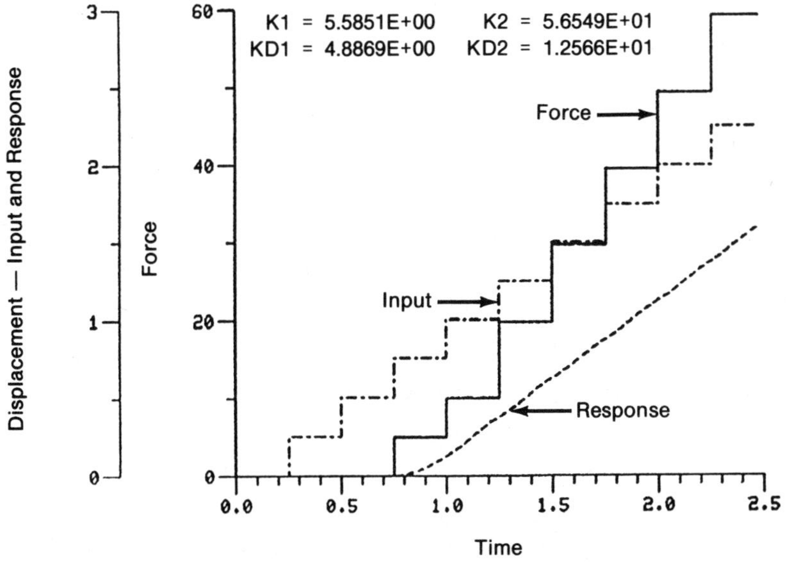

FIGURE 5-18
Response and actuator force for unit ramp input. $\omega_n = 1.0$, $\zeta = 0$, T = 0.25, repeated poles at z = 0.5.

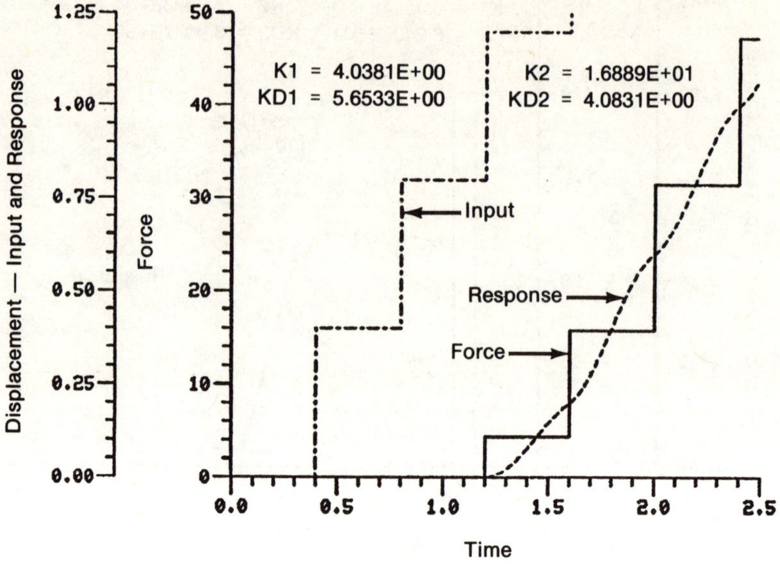

FIGURE 5-19
Response and actuator force for unit ramp input. $\omega_n = 1.0$, $\zeta = 0$, T = 0.4, repeated poles at z = 0.

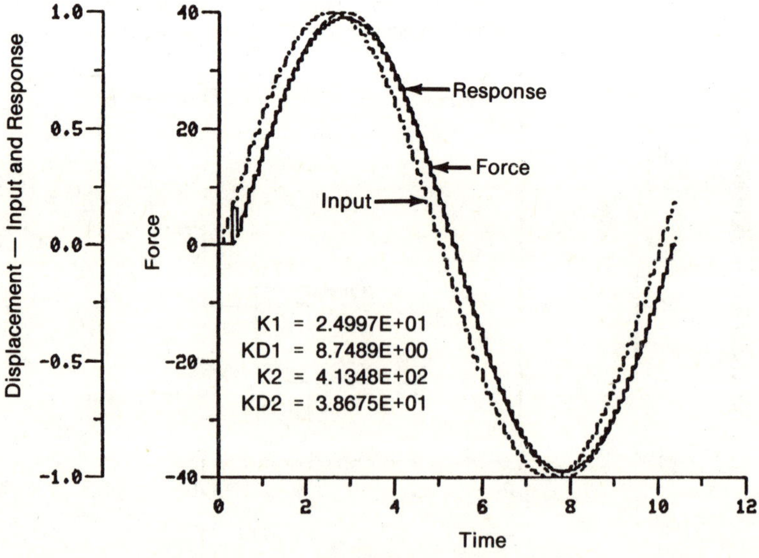

FIGURE 5-20
Response and actuator force for harmonic input, sin $0.2\pi t$. $\omega_n = 1.0$, $\zeta = 0$, T = 0.1, repeated poles at z = 0.

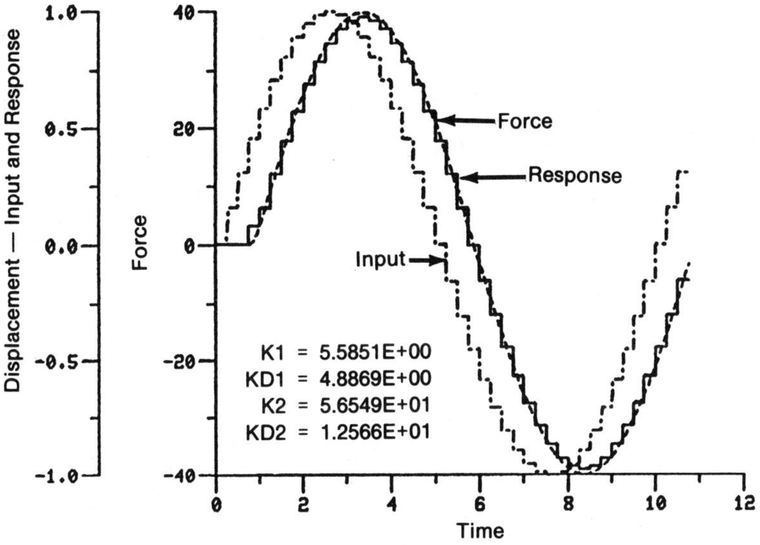

FIGURE 5-21
Response and actuator force for harmonic input, sin $0.2\pi t$. $\omega_n = 1.0$, $\zeta = 0$, T = 0.25, repeated poles at z = 0.

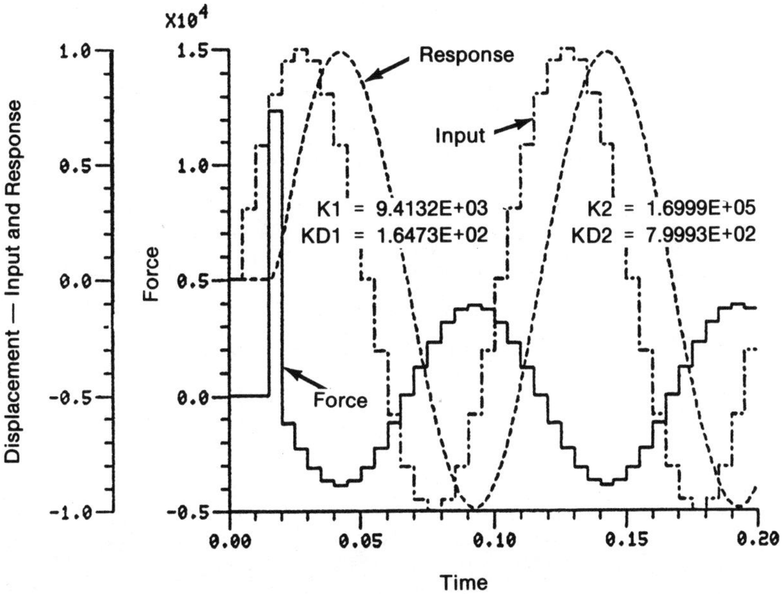

FIGURE 5-22
Response and actuator force for harmonic input, sin $20\pi t$. $\omega_n = 1.0$, $\zeta = 0$, T = .005, repeated poles at z = 0.

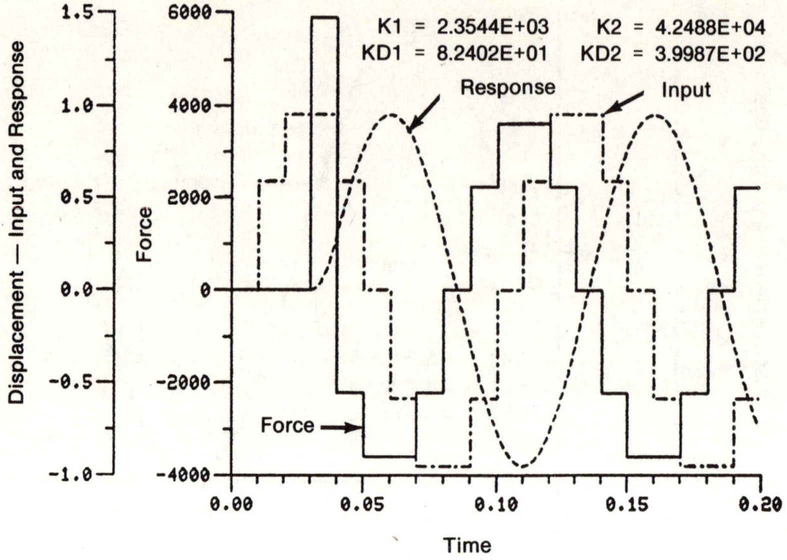

FIGURE 5-23
Response and actuator force for harmonic input, sin $20\pi t$. $\omega_n = 1.0$, $\zeta = 0$, T = 0.01, repeated poles at z = 0.

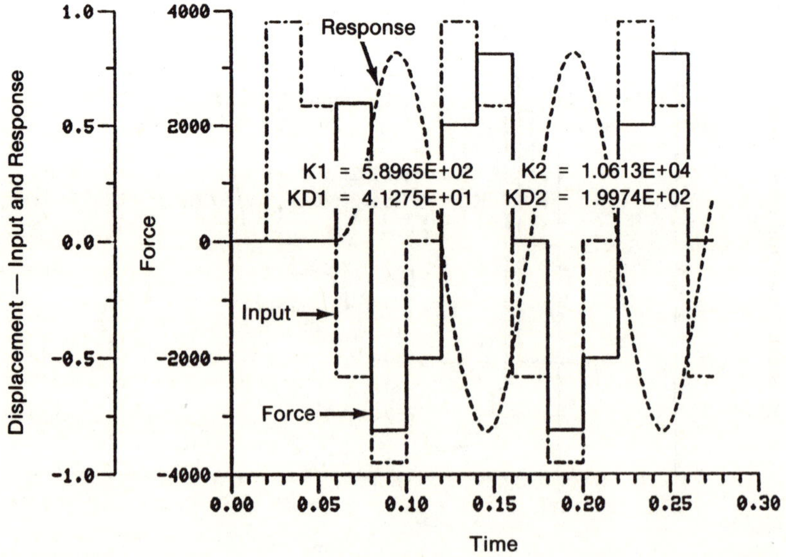

FIGURE 5-24
Response and actuator force for harmonic input, sin $20\pi t$. $\omega_n = 1.0$, $\zeta = 0$, T = 0.02, repeated poles at z = 0.

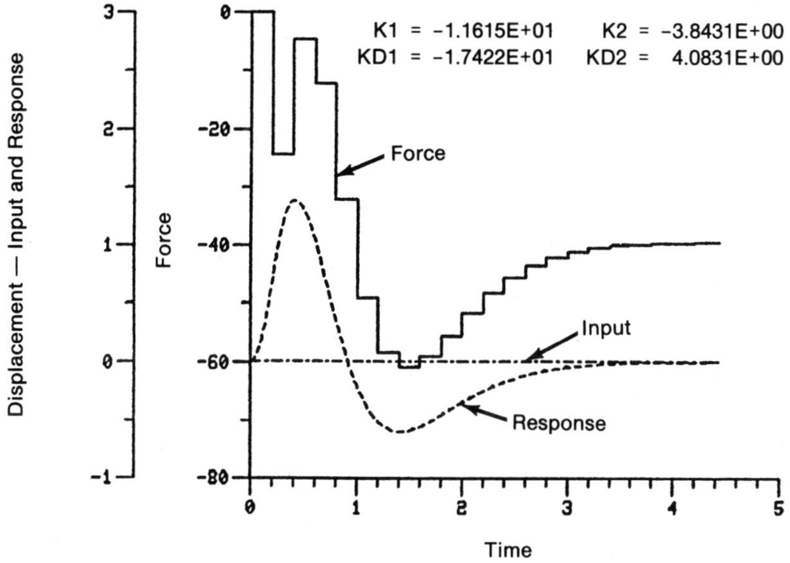

FIGURE 5-25
Response and actuator force for step disturbance of 39.48. $\omega_n = 1.0$, $\zeta = 0$, T = 0.2, repeated poles at z = 0.5.

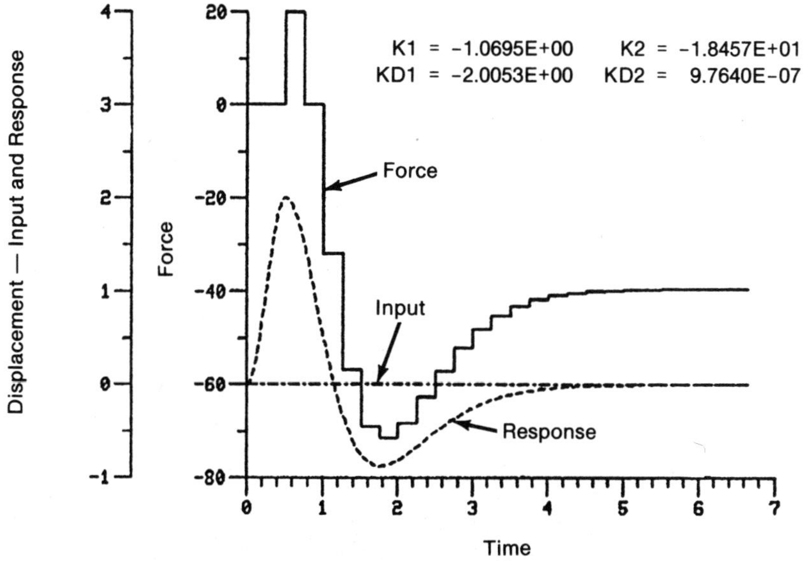

FIGURE 5-26
Response and actuator force for step disturbance of 39.48. $\omega_n = 1.0$, $\zeta = 0$, T = 0.25, repeated poles at z = 0.5.

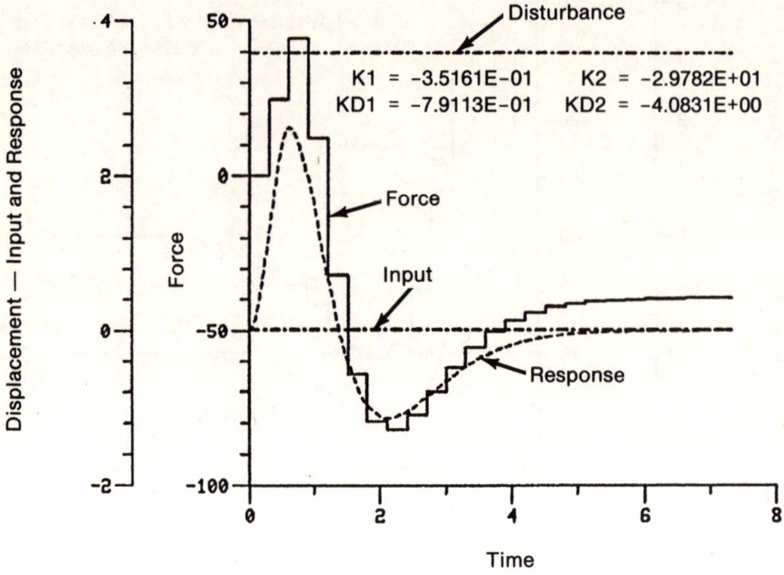

FIGURE 5-27
Response and actuator force for step disturbance of 39.48. $\omega_n = 1.0$, $\zeta = 0$, T = 0.3, repeated poles at z = 0.5.

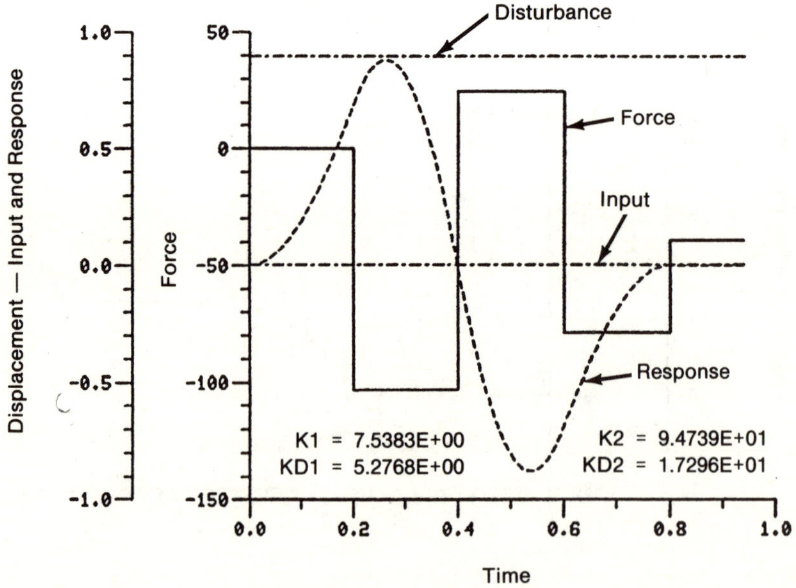

FIGURE 5-28
Response and actuator force for step disturbance of 39.48. $\omega_n = 1.0$, $\zeta = 0$, T = 0.2, repeated poles at z = 0.

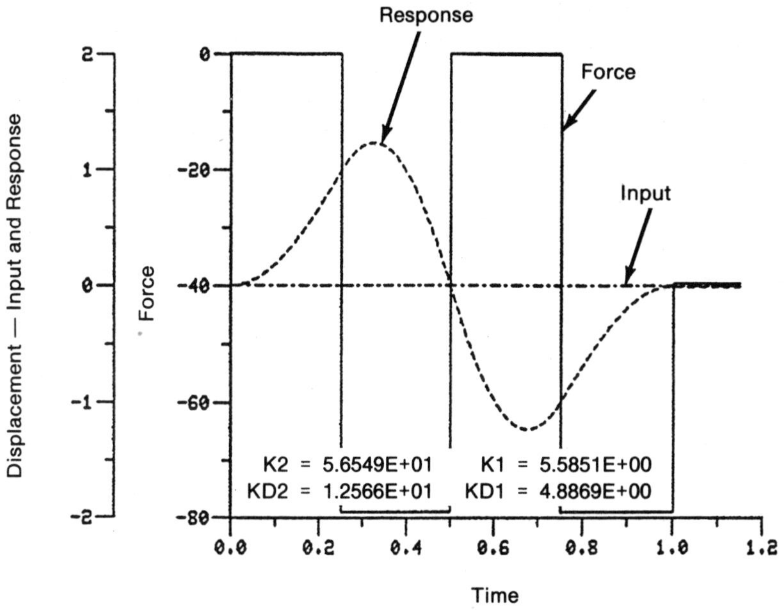

FIGURE 5-29
Response and actuator force for step disturbance of 39.48. $\omega_n = 1.0$, $\zeta = 0$, T = 0.25, repeated poles at z = 0.

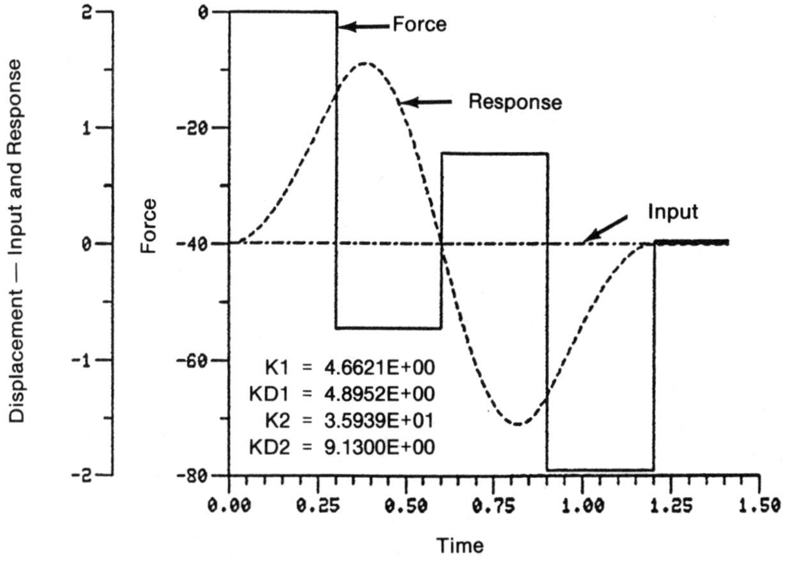

FIGURE 5-30
Response and actuator force for step disturbance of 39.48. $\omega_n = 1.0$, $\zeta = 0$, T = 0.3, repeated poles at z = 0.

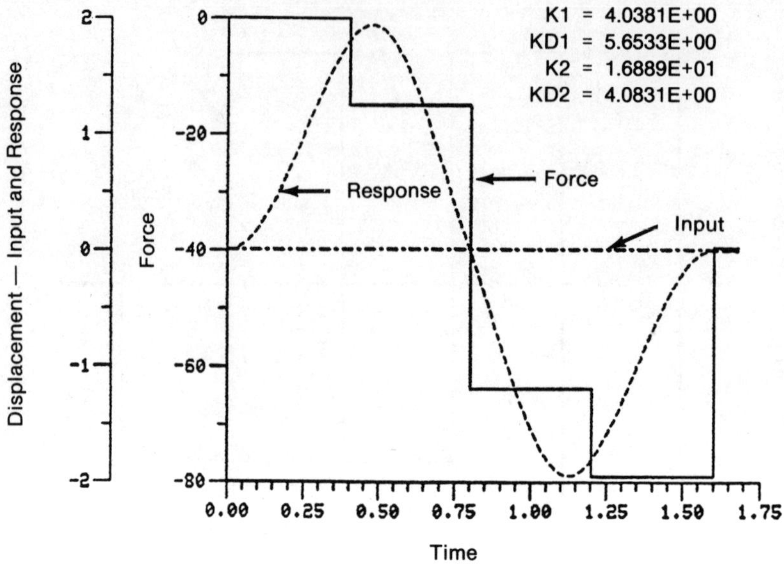

FIGURE 5-31
Response and actuator force for step disturbance of 39.48. $\omega_n = 1.0$, $\zeta = 0$, T = 0.4, repeated poles at z = 0.

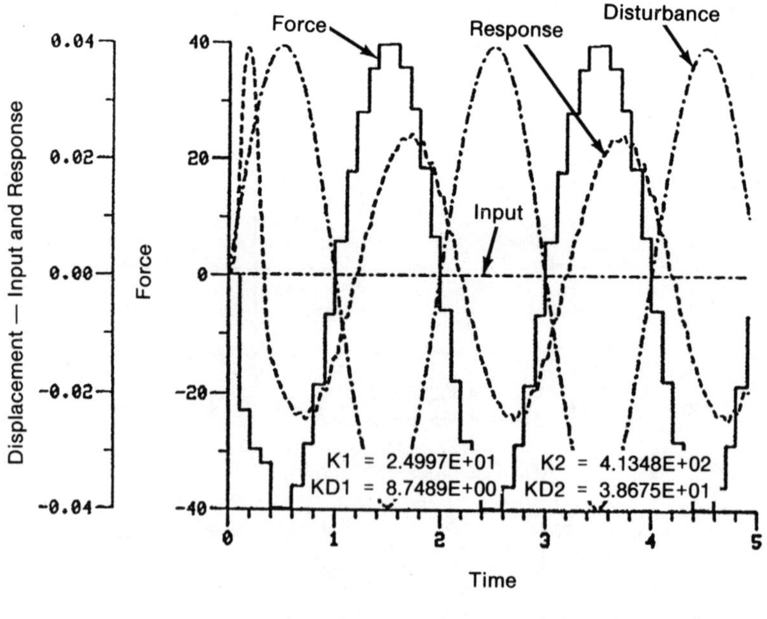

FIGURE 5-32
Response and actuator force for harmonic disturbance force, 39.48 sin πt. $\omega_n = 1.0$, $\zeta = 0$, T = 0.1, repeated poles at z = 0. Input is zero.

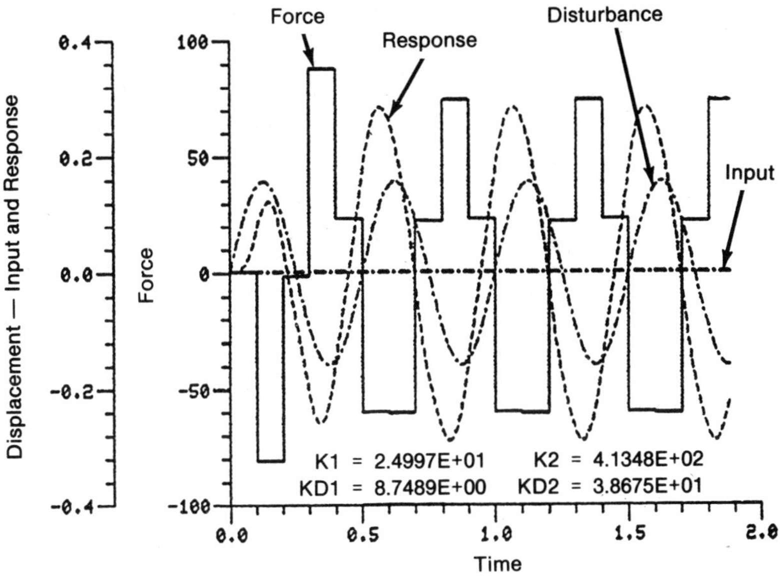

FIGURE 5-33
Response and actuator force for harmonic disturbance force, 39.48 sin $4\pi t$. $\omega_n = 1.0$, $\zeta = 0$, $T = 0.1$, repeated poles at $z = 0$. Input is zero.

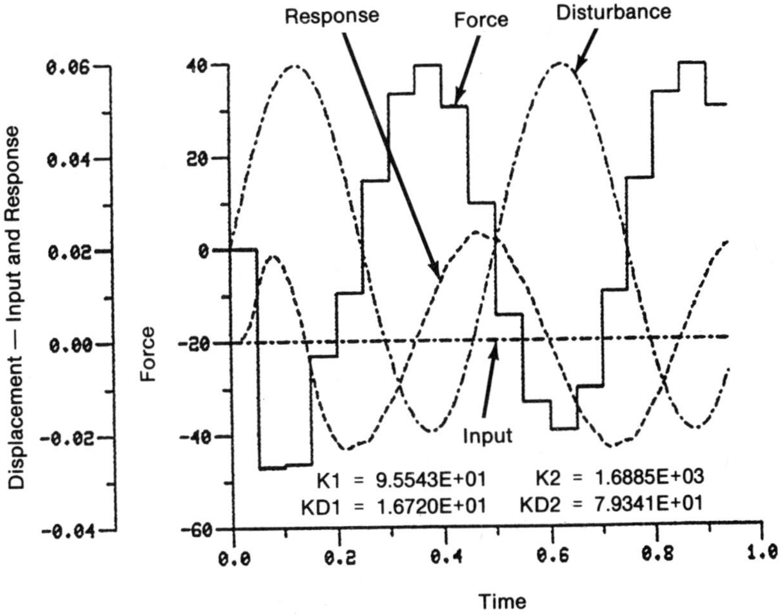

FIGURE 5-34
Response and actuator force for harmonic disturbance force, 39.48 sin $4\pi t$. $\omega_n = 1.0$, $\zeta = 0$, $T = 0.05$, repeated poles at $z = 0$. Input is zero.

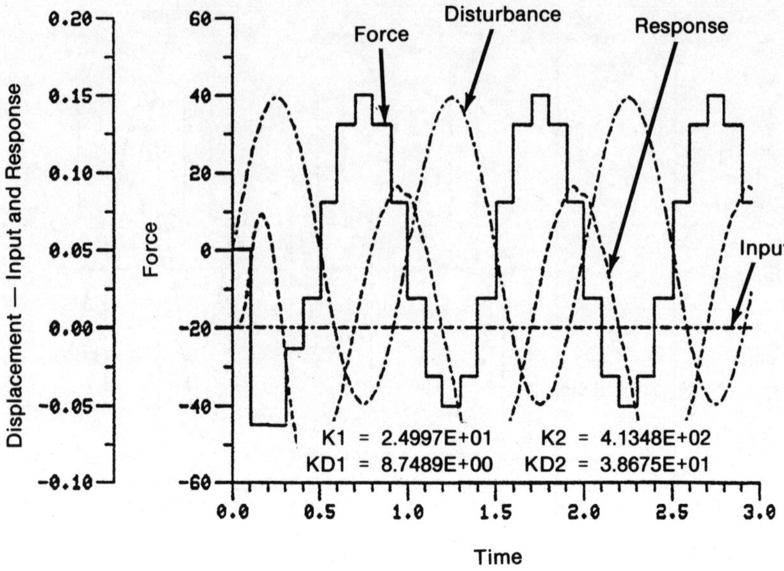

FIGURE 5-35
Response and actuator force for harmonic disturbance force, 39.48 sin $2\pi t$. ω_n = 1.0, ζ = 0, T = 0.1, repeated poles at z = 0. Input is zero.

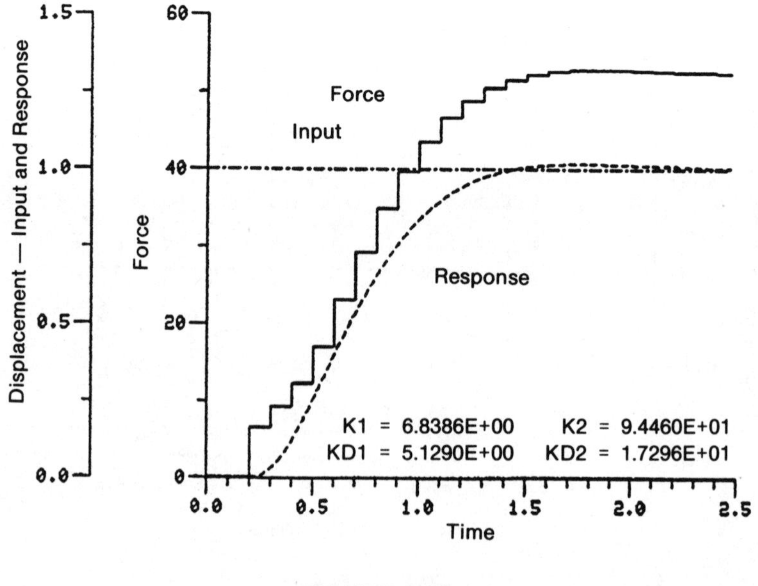

FIGURE 5-36
Response and actuator force for step input. Model ω_n = 1.0, actual ω_n = 1.15, ζ = 0, T = 0.1, repeated poles at z = 0.5.

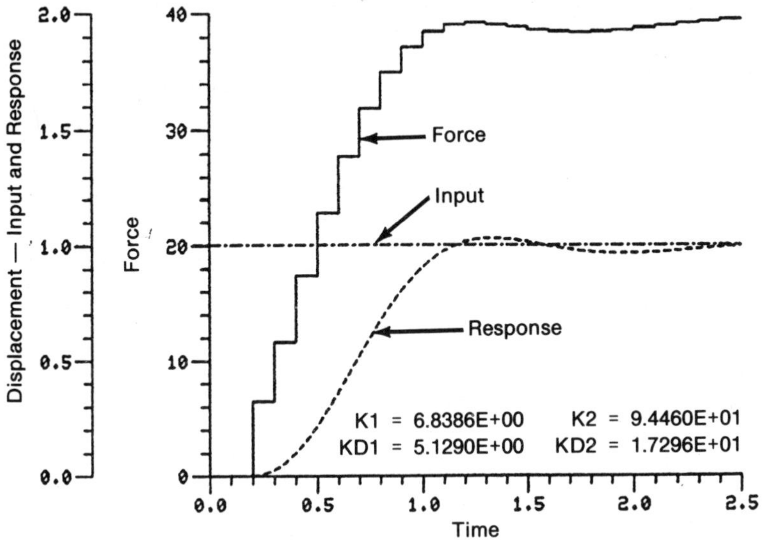

FIGURE 5-37
Response and actuator force for step input. $\omega_n = 1.0$, model $\zeta = 0$, actual $\zeta = 0.5$, $T = 0.1$, repeated poles at $z = 0.5$.

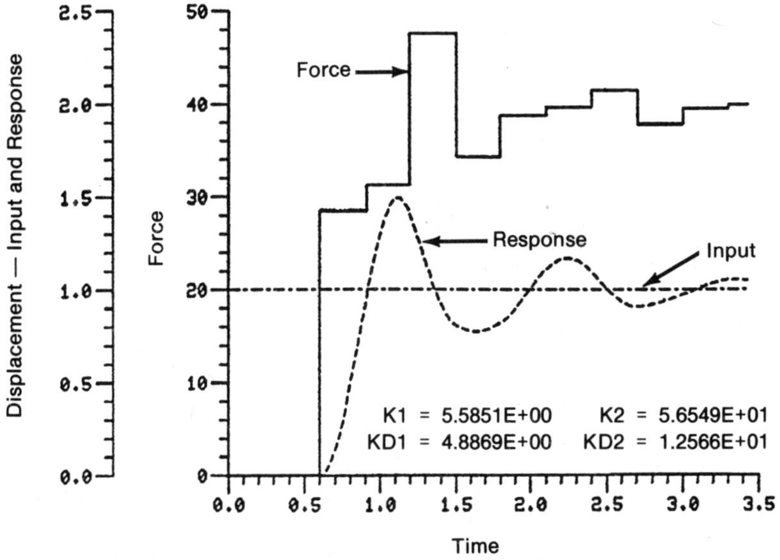

FIGURE 5-38
Response and actuator force for step input. $\omega_n = 1.0$, $\zeta = 0$, model $T = 0.25$, actual $T = 0.3$, repeated poles at $z = 0$.

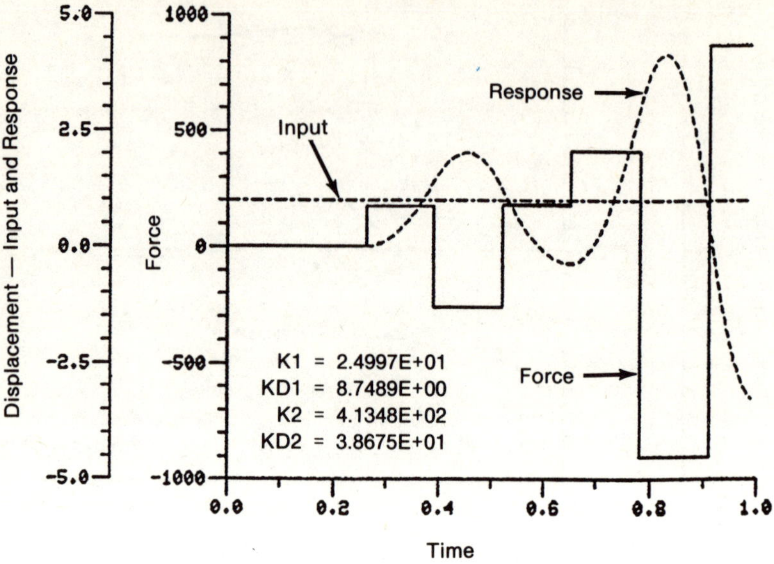

FIGURE 5-39
Response and actuator force for step input. $\omega_n = 1.0$, $\zeta = 0$, model T = 0.1, actual T = 0.13, repeated poles at z = 0.

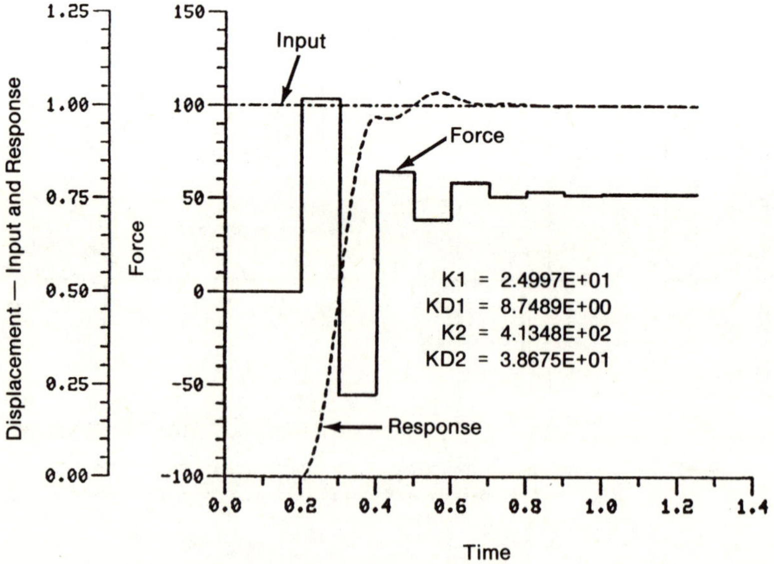

FIGURE 5-40
Response and actuator force for step input. Model ω_n = 1.0, actual $\omega_n = 1.15$, $\zeta = 0$, T = 0.1, repeated poles at z = 0.

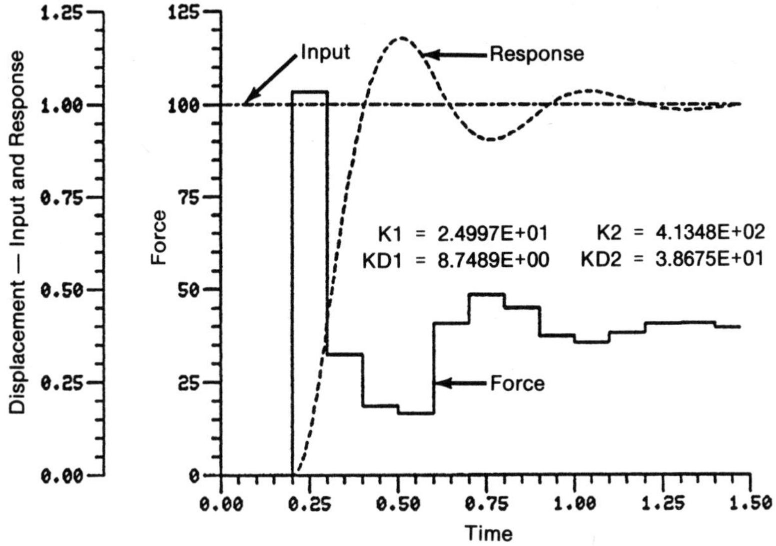

FIGURE 5-41
Response and actuator force for step input. $\omega_n = 1.0$, model $\zeta = 0$, actual $\zeta = 0.5$, T = 0.1, repeated poles at z = 0.

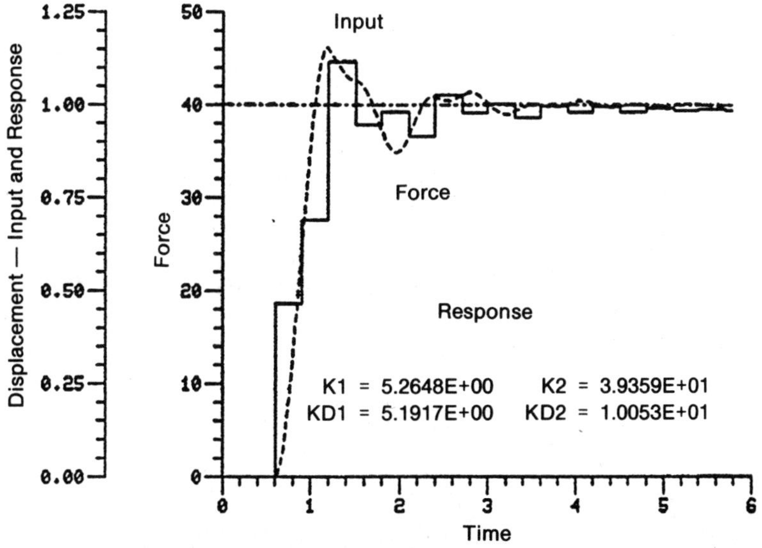

FIGURE 5-42
Response and actuator force for step input. $\omega_n = 1.0$, $\zeta = 0$, model T = 0.25, actual T = 0.3, repeated poles at z = 0.1.

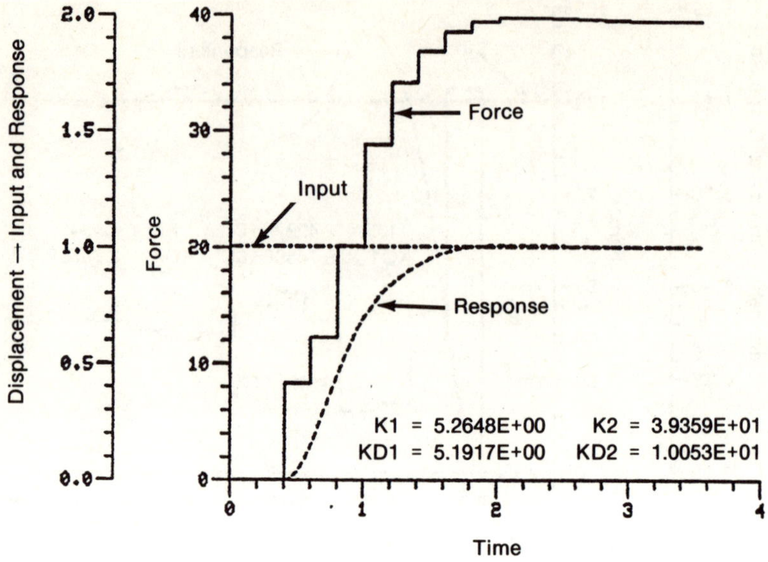

FIGURE 5-43
Response and actuator force for step input. ω_n = 1.0, ζ = 0, model T = 0.25, actual T = 0.2, repeated poles at z = 0.1.

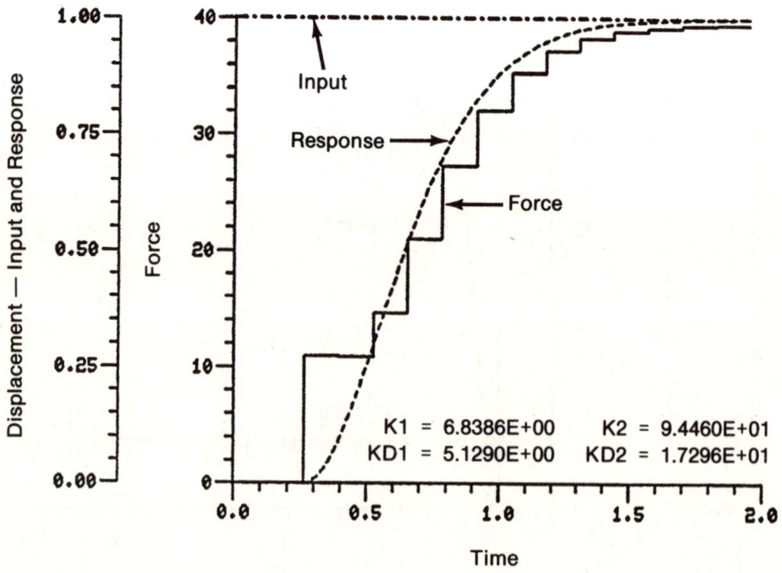

FIGURE 5-44
Response and actuator force for step input. ω_n = 1.0, ζ = 0, model T = 0.1, actual T = 0.13, repeated poles at z = 0.5.

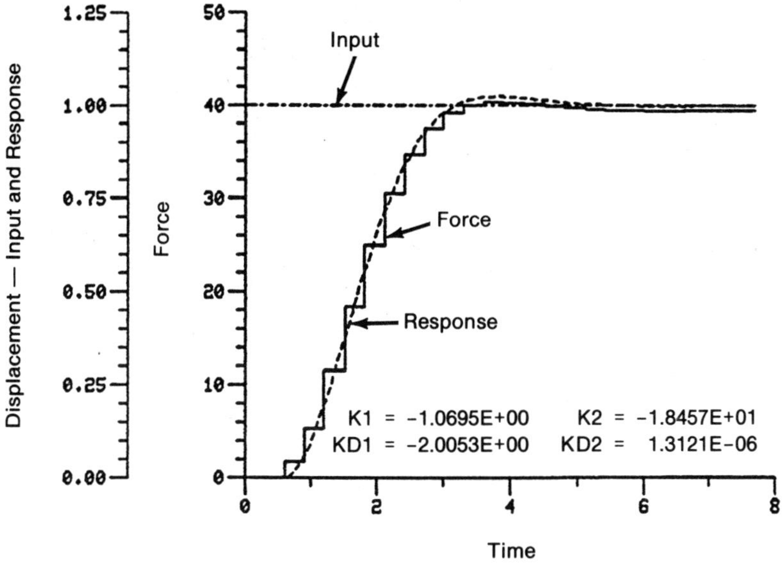

FIGURE 5-45
Response and actuator force for step input. ω_n = 1.0, ζ = 0, model T = 0.25, actual T = 0.3, repeated poles at z = 0.5.

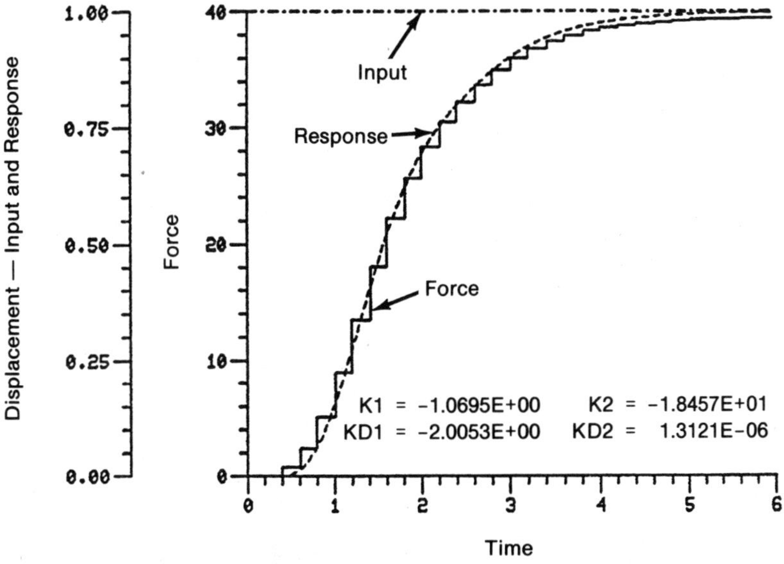

FIGURE 5-46
Response and actuator force for step input. ω_n = 1.0, ζ = 0, model T = 0.25, actual T = 0.2, repeated poles at z = 0.5.

Conclusion

A FORTRAN listing of the program for which simulations are presented is shown in the Appendix as DIG.

CHAPTER VI

The Reduced-Order Error Observer

Synthesis and analysis techniques for the design of multivariable feedback control systems for time response generally include the linear quadratic regulator (LQR) and the pole assignment approach. Both methods depend upon complete state variable feedback information, which may not always be available. Inaccessible state variables generally are reconstructed or synthesized with a *state observer* that uses information received from measurements of the input and the available output of the system. The observer also requires a model of the system in the reconstruction or synthesis of these inaccessible states.

The observer may be a full-order type that generates estimates of all state variables including those that are measurable. A reduced-order observer can be implemented to generate only those state variables that are inaccessible. The full-order observer can be implemented readily, but it does have the drawback that the entire state must be computed. In digital systems this could be critical, because the time required to compute these states could result in significant time delays. The reduced-order observer, on the other hand, generates only those state variables that are needed. Its implementation for large systems, however, requires a considerable amount of matrix manipulations because the problem is put into the adjoint phase-variable canonical form (1).

The required successive state variable matrix transformations may require considerable computing time for implementation and produce time delays. The full-order and the reduced-order state observer each depend upon a system model for estimates of the state variables.

It is possible also to carry out limited pole assignment in systems with incomplete state feedback. One pole each can be assigned for each state measured (2). The remaining poles that cannot be placed by the analyst may be in the vicinity of those assigned. Such a guarantee, however, is not available. In this chapter, we consider an approach that allows pole assignment with incomplete state feedback information but without the use of either the traditional full-order or reduced-order observers. Moreover, matrix transformations are eliminated.

The method is unlike any in the author's experience. It is similar to the reduced-order state observer in that the order of the system is increased by one for each inaccessible state. The similarity stops here, however, because this new device becomes an integral part of the control system. Unlike the state observer, it is not an adjunct that provides information to the system. It is possible to show that, with the method, some measure of the error associated with the inaccessible state(s) is synthesized. In a design setting, the choice of the asymptotic state observer or the error observer for inaccessible states depends upon the system and its design specifications.

The discovery of this observer has revealed another relationship between classical control theory and modern control theory. This will be illustrated in this chapter by considering a linear servomechanism with incomplete state feedback and constant feedback gains. A relationship will be established between the classical block diagram and the optimal linear quadratic servomechanism (LQS).

Objective

The objective of the work in this chapter is to construct and evaluate a feedback control algorithm that gives complete pole assignment capability with only partial state variable feedback information but does not use a state observer.

Implementation of this effort will be a significant contribution to the synthesis and design of feedback controllers for very large systems such as machine tools with in-process gaging and control, internal combustion engines, space structures, and multi-degree-of-freedom robots. Active control of bearing supports for rotating machinery is another important application. In these large systems, several of the state variables are usually inaccessible. In particular, velocity information may be difficult to obtain. Also, because models of these systems are of very high order, pole assignment has required much laborious and tedious effort. Most of this effort and the attendent opportunities for error can be eliminated with a symbolic and algebraic manipulation language.

Error Observer

In this development we describe the error observer first for a low-order continuous system because most of us have a better feeling for such systems and can use our backgrounds and insights to help with understanding. Application to a digital or sampled-data system follows naturally. Low-order systems facilitate insight, but they also suggest the extension to higher-order systems, which are treated in a subsequent section. We consider a second-order plant such as a linear spring-mass system. This example is significant because one of the states could be the velocity. Velocities often are difficult to measure and, hence, may be inaccessible. In Figure 6-1, the block diagram

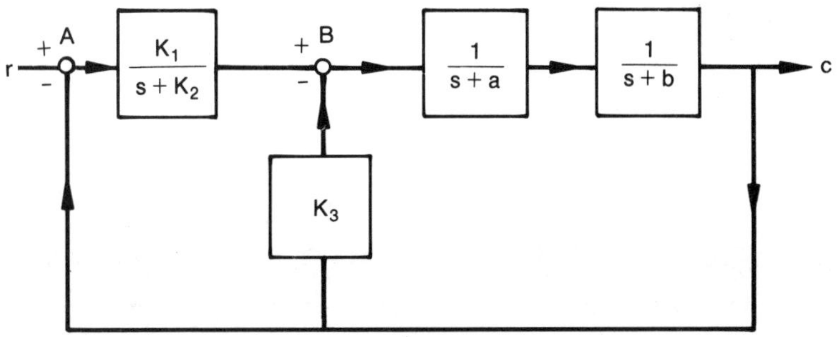

FIGURE 6-1
"Error observer" concept.

illustrates the concept of "error observer" algorithm. The output of point A is clearly the error associated with position. The output of point B can be thought of as some measure of the error associated with the velocity. This association can be given some validity by considering the block diagram for the second-order linear quadratic servomechanism shown in Figure 6-2. Complete state feedback is assumed, so an observer is not required. Here the output of A is the error associated with position as in Figure 6-1, but the output of B is seen to be closely related to the error associated with velocity or $\dot{c}$. The transfer function for the system of Figure 6-1 is

$$\frac{c}{r} = \frac{K_1}{s^3 + s^2(K_2+a) + s(K_3+b+aK_2) + (K_1+bK_2+K_2+K_3)} \tag{6-1}$$

Complete pole assignment is possible because K_1, K_2, and K_3 can be determined to set the coefficients of s^0, s^1, and s^2 to any specified value. In a similar fashion, it can be shown that complete pole assignment is possible for the configuration in Figure 6-2. The sampled-data digital controller version of Figure 6-1 is illustrated in Figure 6-3.

For the systems in Figures 6-1 and 6-3, the steady-state error for a step input is not zero. This error can be eliminated by adding a stage of integral compensation as shown in Figure 6-4 for the continuous version and in Figure 6-5 for the discrete

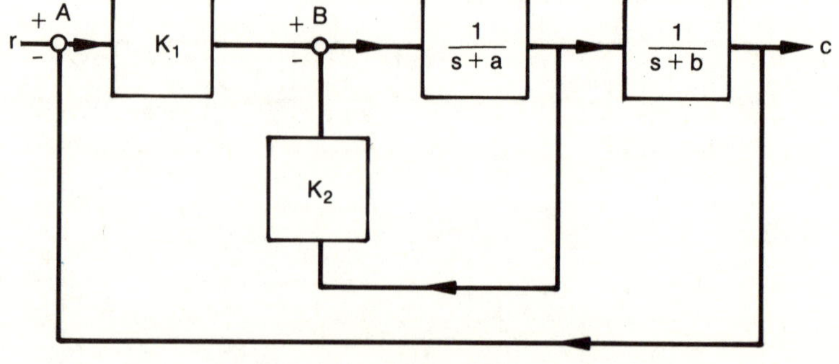

FIGURE 6-2
Linear quadratic servomechanism.

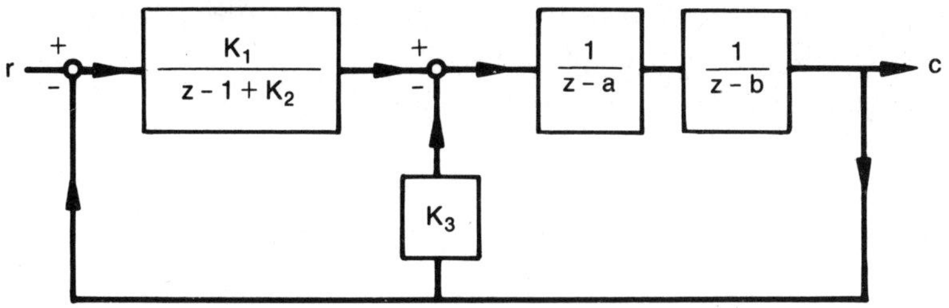

FIGURE 6-3
Digital equivalent of Figure 6-1.

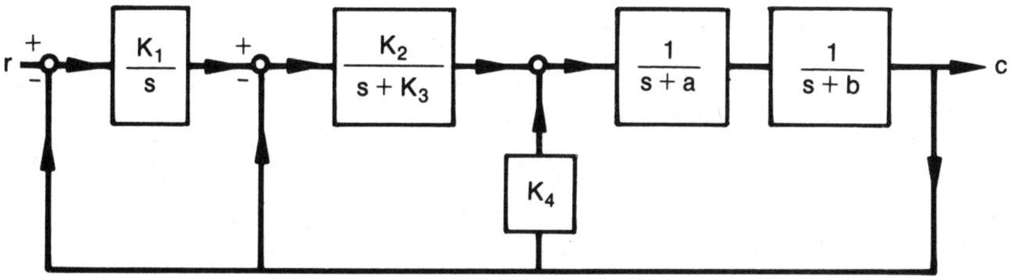

FIGURE 6-4
Plant with error observer and one stage of integral compensation.

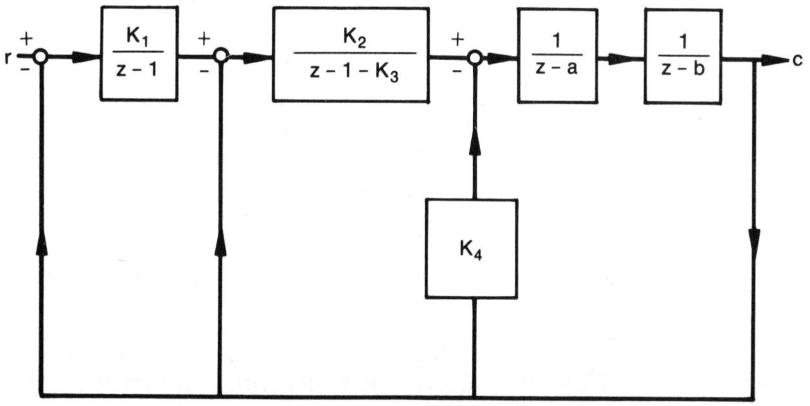

FIGURE 6-5
Sampled-data equivalent of Figure 6-4.

system. The constants in Figures 6-3 and 6-5 differ from those in Figures 6-2 and 6-4, respectively.

The transfer function for the system in Figure 6-4 is

$$\frac{c}{r} = \frac{K_1 K_2}{s^4 + s^3(a+b+K_3) + s^2(K_4+ab+K_3a+K_3b) + s(K_2+K_3K_4+abK_3) + K_1K_2}$$

Again, pole assignment is possible because the coefficients of s^0, s^1, s^2, and s^3 can be set to any specified value. The order of computation would be to obtain K_3 followed by K_4, K_2, and K_1. The algebraic operations required here are direct. For the sampled data system, however, the work is burdensome enough that REDUCE™ was used to solve for the gains algebraically. Higher-order systems would use some variation of the linear quadratic servomechanism or regulator to obtain the gains. Simulations, similar to those of the previous chapter, were made with generally expected results. This system did not show the same robustness, however, that we observed in Chapter 5. We believed that this followed from the use of only one stage of IESF compensation. In an attempt to improve robustness, therefore, two such stages were used with the error observer. Performance degraded further, however, compared to the single stage of EISF compensation. This result was totally unexpected and could be a starting point for further research.

Extension to Higher-Order Systems

As the order of the system increases there will be some question about the implementation of the error observer, i.e., what "error" must the observer synthesize? The method to be described here is interesting in that it shows the previously mentioned relationship between modern control and classical control. This new connection is described best by considering the chronological steps that led to its formulation. The starting point is a plant with a number of first-order poles in series and in complete state feedback. This recalls the linear quadratic servomechanism with constant feedback gains and complete state feedback. The question is whether the error observer can be employed, and, if so, how can it be implemented? The vector/matrix block diagram of the control system, Figure 6-6, does not suggest any way to realize the implementation. If, however, we begin with a first-order pole with feedback, Figure 6-7, a sequence can be developed to show how to use the "error observer" where there is incomplete state feedback.

In the next step, we consider a plant with a second-order pole (or two first-order poles in series) with only final output feedback, Figure 6-8. The error observer follows from the previous discussion and Figure 6-1. Pole assignment is a direct process here. When the poles have been specified, the coefficients of power of s can be computed which allows K_2, K_3, and K_1 to be determined in that order.

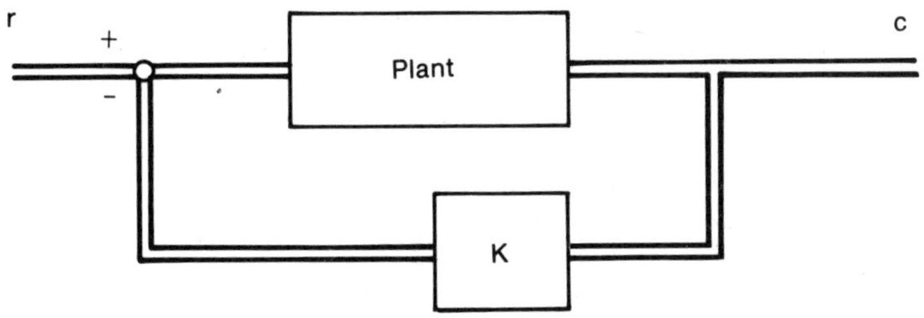

FIGURE 6-6
Vector/matrix block diagram for multivariable system.

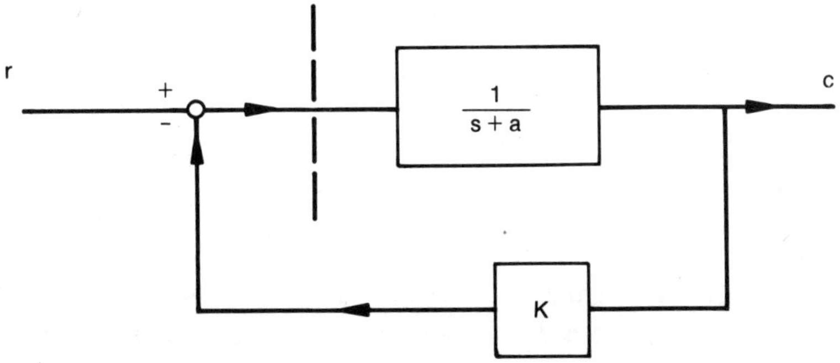

FIGURE 6-7
First-order pole with state feedback.

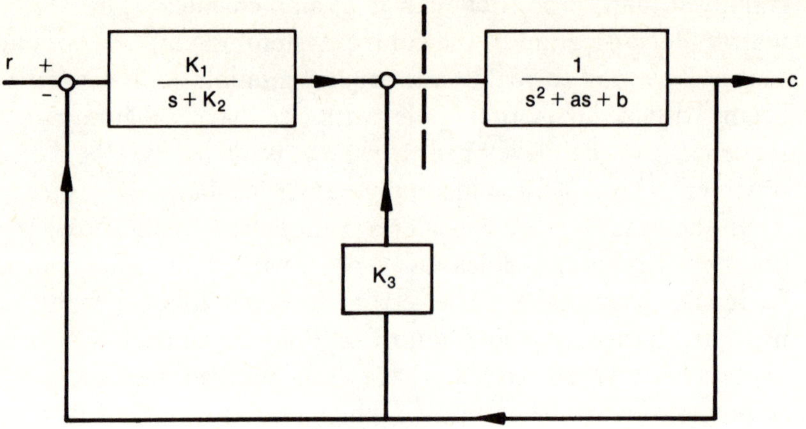

FIGURE 6-8
Second-order plant with an error observer.

The overall transfer function is

$$\frac{c}{r} = \frac{K_1}{s^3 + s^2(K_2+a)+s(K_2\mathrm{a}+K_3+b) + K_1+bK_2 + K_2K_3}$$

In the next step we consider a three-pole plant with final output only, just as in the previous situation. The complete control system with two "error observers" is shown in Figure 6-9. Pole assignment is feasible here also but with one possible difficulty. The coefficients of powers of s are nonlinear so that for certain combinations of a, b, d, and pole locations, the K_i may be found to be complex. Further experience is required here. The block diagram here and that of Figure 6-1 establish the pattern for higher-order plants with only output feedback.

The control system for a three-pole plant with only one state feedback missing may take on either of two forms. The first is shown in Figure 6-10 where feedback is available from the last two states. Specification of the poles here permits direct calculation of K_3, K_4, K_2, and K_1 in that order.

The second form is shown in Figure 6-11 where feedback is unavailable for the output of the second pole. Pole assignment is feasible here also but with the possible difficulty of complex K_i's because the gains appear nonlinearly, just as in Figure 6-9 for the first three-pole plant configuration.

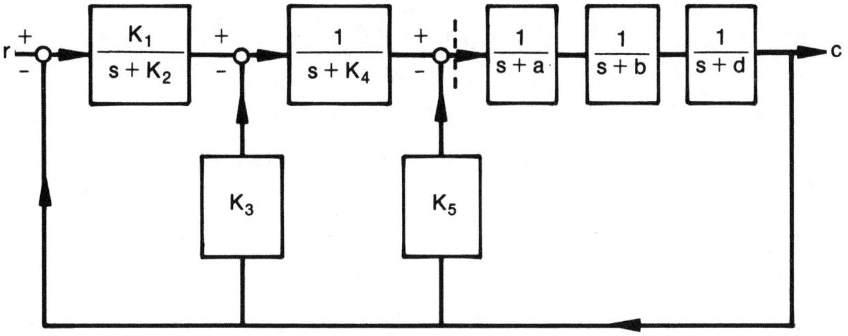

$$\frac{c}{r} = K_1\{s^5 + s^4(a+b+d+K_2+K_4) + s^3[ab+ad+bd+(a+b+d)(K_2+K_4) + K_2K_4] + s^2[K_5+abd+K_2K_4(a+b+d) + (K_2+K_4)(ab+ad+bd)] + s[K_3+K_5(K_4+K_2) + abd(K_2+K_4) + K_2K_4(ab+da+bd)] + K_1 + K_2K_3 + K_2K_4K_5 + abdK_2K_4\}^{-1}$$

FIGURE 6-9
Three-pole plant and controller with final output feedback only.

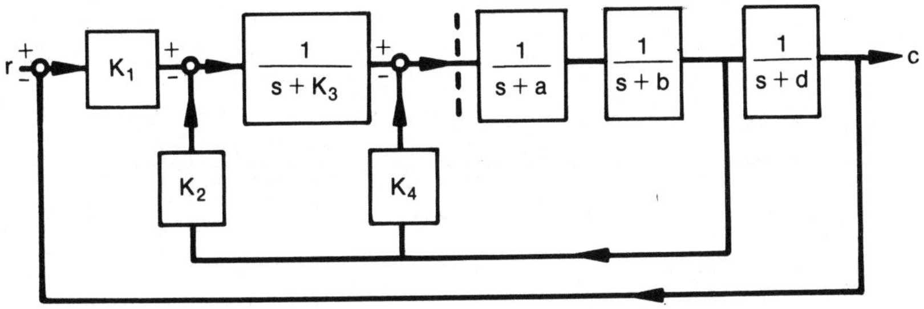

$$\frac{c}{r} = K_1\{s^4+s^3(b+d+a+K_3) + s^2(K_4+bd+aK_3+ab+ad+K_3b+K_3d) + s(K_2+K_4d+K_3K_4+abd+K_3bd+abK_3+adK_3) + K_1+K_2d\}^{-1}$$

FIGURE 6-10
Three-pole plant with controller—feedback from two states.

In Figures 6-7 through 6-11 the basic building blocks are given. Also in these figures a vertical dashed line is shown in the forward path. Each diagram can be broken at the dashed line and one of the basic building blocks can be inserted here. This process can be repeated until any system of interest is constructed.

The last step in the process of establishing this bridge between classical control and modern control is to consider a three-pole plant with complete state feedback and construct the block diagram using the basic building block of Figure 6-7. This results in the cascaded control system shown in Figure 6-12. This will be

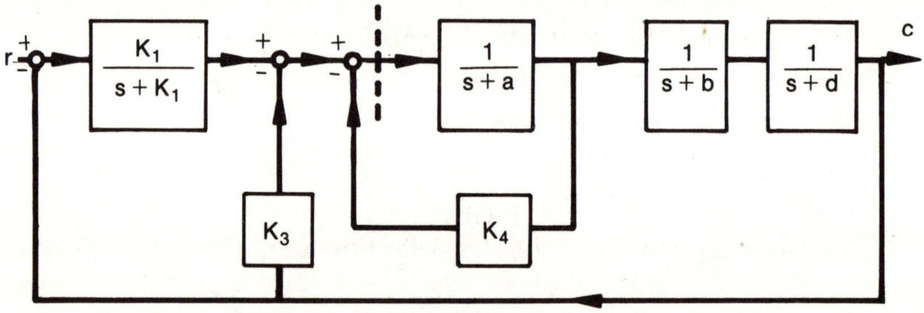

$$\frac{c}{r} = K_1\{s^4 + s^3(K_2+K_4+a+b+d) + s^2(K_2K_4+bK_4+dK_4+bd+aK_2+ab+ad + K_2b+K_2d) + s(K_3+bdK_4+bK_2K_4+dK_2K_4+abd+K_2bd + abK_2 + adK_2) + K_1 + K_2K_3 + bdK_2K_4 + abdK_2\}^{-1}$$

FIGURE 6-11
Three-pole plant with controller—feedback unavailable from second pole.

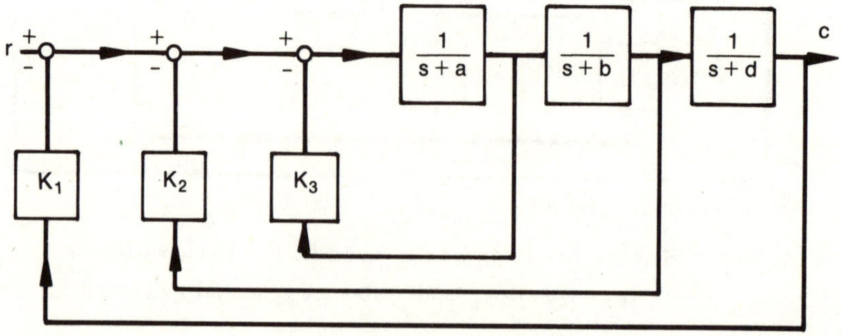

FIGURE 6-12
Cascaded control system.

recognized as the equivalent of Figure 6-6 for the linear quadratic servomechanism with constant feedback gains. By way of the "error observer" and classical block diagram representations, we have constructed a system usually associated with modern control.

Finally, in answer to the question of how to implement the "error observer" in a large system, the building block approach will provide the correct error compensation at the correct place in the controller.

The FORTRAN implementation of the error observer is included in DIG in the Appendix.

References

1. B. C. Kuo, *Digital Control Systems*, Holt, Rinehart, and Winston, New York, 1980.
2. Davison, E. J., "On Pole Assignment in Linear Systems with Incomplete State Feedback," *IEEE Trans. on Automatic Control*, Vol. AC-15, June 1980, pp. 348–351.

Appendix

CONTROL

```
      IMPLICIT REAL*8(A-H,O-Y)
      EXTERNAL FCT,OUTP
      DIMENSION Y(6),DERY(6),PRMT(5),AUX(8,6),XK3(3),XKD3(3),XK2(3),
     1XKD2(3),R(3),ZX(600),ZY4(600),ZY5(600),ZY6(600),ZI(600)
      COMMON XK1,S1,A1,XKD1,XK2,S2,A2,XKD2,XK3,S3,A3,XKD3,ROL,XLA,BKLA
     1BOJ,XKOJ,U1,U2,U3,XL1,XL2,XL3,Y61,S12,S22,S32,XJ,RA,B,XKB,XKI,R,
     2ZX,ZY4,ZY5,ZY6,ZI,C1,C2,YR61
      COMMON/II/IP,JS,JC,KC,MC,NS,MPRINT,IC12
      DO 453 I=1,100
      ZX(I)=0.
      ZY4(I)=0.
      ZY5(I)=0.
  453 ZY6(I)=0.
      IC12=0
      MC=0
      KC=1
      IP=0
C234567
      WRITE(6,*)' ENTER STEP SIZE H'
      READ(5,*)PRMT(3)
      WRITE(6,*)' STEP SIZE H=',PRMT(3)
      WRITE(6,*)'ENTER TWO INPUT COEFFICIENTS'
      READ(5,*)Y61,YR61
      WRITE(6,*)' STEP INPUT=',Y61,'RAMP INPUT=',YR61
      WRITE(6,*)' ENTER TORQUE DISTURBANCE COEFFICIENTS C1 AND C2
     1; C1 SHOULD BE LESS THAN KI*U2. C1 IS STEP, C2 IS RAMP RATE
     2; ALSO ENTER 1 FOR C1 OR C2 NOT ZERO'
      READ(5,*)C1,C2,IC12
      WRITE(6,*)'C1=',C1,'C2=',C2,'IC12=',IC12
      WRITE(6,*)' ENTER R(1)'
      READ(5,*)R(1)
      WRITE(6,*)' R(1)=',R(1)
      WRITE(6,*)' ENTER U1 AND U2'
      READ(5,*)U1,U2
```

```
      WRITE(6,*)' U1=',U1
      WRITE(6,*)' U2=',U2
      WRITE(6,*)' IF U3 IS ON, ENTER 1'
      READ(5,*)MU3
      WRITE(6,*)' MU3=',MU3
      WRITE(6,*)' ENTER NP AND NS'
      READ(5,*)NP,NS
      WRITE(6,*)' NP=',NP,' NS=',NS
      WRITE(6,*)' ENTER MPRINT'
      READ (5,*)MPRINT
      WRITE(6,*)' MPRINT=',MPRINT
      JC=NS-1
      NDIM=6
      S22=1.D0
      S32=1.D0
      S12=1.D0
      XJ=15.D-4
      XKB=.02125D0
      XLA=.005D0
      BKLA=4.25D0
      BOJ=40.D0
      XKOJ=2000.D0
      ROL=1000.D0
      XKI=3.D0
      B=.06D0
      RA=5.D0
      S1=1.D0
      S2=1.D0
      S3=1.D0
      A1=0.D0
      A2=0.D0
      A3=0.D0
      R2=R(1)*R(1)
      JS=1
      XKD3(1)=6.D0*R(1)*XLA-RA-XLA*B/XJ
      XK3(1)=15.D0*R2*XLA-(XKD3(1)*B+XKB*XKI+B*RA)/XJ
      XKD2(1)=20.D0*R2*R(1)*XLA*XJ/(XK3(1)*XKI)-B/XKI
      XK2(1)=15.D0*R2*R2*XLA*XJ/(XKI*XK3(1))
      XKD1=6.D0*R2*R2*R(1)*XLA*XJ/(XKI*XK3(1)*XK2(1))
      XK1=XLA*XJ*R2*R2*R2/(XK2(1)*XK3(1)*XKI)
      R(3)=(RA+XKD3(1))/(2.D0*XLA)
      XKD3(3)=2.D0*R(3)*XLA-RA
      R(2)=(XKD3(1)+RA+XLA*B/XJ)/(4.D0*XLA)
      R2=R(2)*R(2)
      XKD3(2)=-RA-XLA*B/XJ+4.D0*R(2)*XLA
      XK3(2)=6.D0*R2*XLA-(B*XKD3(2)+B*RA+XKI*XKB)/XJ
      XKD2(2)=(4.D0*R2*R(2)*XLA*XJ/XK3(2)-B)/XKI
      XK2(2)=R2*R2*XLA*XJ/(XK3(2)*XKI)
      XK3(3)=R(3)*R(3)*XLA
      XKD3(3)=2.D0*R(3)*XLA-RA
      XK2(3)=XK2(2)
      XKD2(3)=XKD2(2)
      WRITE(6,*)R
      WRITE(6,201)XK1
      WRITE(6,201)XKD1
      WRITE(6,200)XK2
      WRITE(6,200)XKD2
      WRITE(6,200)XK3
      WRITE(6,200)XKD3
  201 FORMAT(D26.16)
  200 FORMAT(3D26.16)
      DIM=NDIM
      XN=1.D0/DIM
      DO 11 I=1,NDIM
      Y(I)=0.D0
   11 DERY(I)=XN
      PRMT(1)=0.D0
      PRMT(2)=100.D0
      PRMT(4)=1.D-6
      U3=1.D20
      IF(MU3.EQ.1)U3=5.D0*U2+.02125D0*U1
```

```
      XL1=-U1
      XL2=-U2
      XL3=-U3
   16 DO 10 I=1,NDIM
   10 DERY(I)=XN
      CALL DKGS(PRMT,Y,DERY,NDIM,IHLF,FCT,OUTP,AUX)
      IF(PRMT(5).EQ.1.DO)GO TO 16
  499 FORMAT(4E13.4)
      CALL INITT(960)
      CALL TERM(2,1024)
      CALL CZAXIS(1)
      CALL BINITT
      CALL SLIMX(353,983)
      CALL NPTS(NP)
      CALL LINE(0)
      CALL XFRM(2)
      CALL YFRM(2)
      CALL CHECK(ZX,ZY6)
      CALL DSPLAY(ZX,ZY6)
      CALL DINITY
      CALL DINITX
      CALL LINE(4)
      CALL XFRM(2)
      CALL YFRM(2)
      CALL CPLOT(ZX,ZI)
      CALL DINITY
      CALL DINITX
      CALL YLOC(-90)
      CALL LINE(3)
      CALL XFRM(2)
      CALL YFRM(2)
      CALL CHECK(ZX,ZY5)
      CALL DSPLAY(ZX,ZY5)
      CALL DINITY
      CALL DINITX
      CALL YLOC(-180)
      CALL LINE(2)
      CALL XFRM(2)
      CALL YFRM(2)
      CALL CHECK(ZX,ZY4)
      CALL DSPLAY(ZX,ZY4)
      CALL TINPUT(I)
      CALL FINITT(0,700)
      STOP
      END
      SUBROUTINE FCT(X,Y,DERY)
      IMPLICIT REAL*8(A-H,O-Y)
      DIMENSION Y(6),DERY(6),XK2(3),XKD2(3),XK3(3),XKD3(3),R(3),
     1ZX(600),ZY4(600),ZY5(600),ZY6(600),ZI(600)
      COMMON XK1,S1,A1,XKD1,XK2,S2,A2,XKD2,XK3,S3,A3,XKD3,ROL,XLA,BKLA,
     1BOJ,XKOJ,U1,U2,U3,XL1,XL2,XL3,Y61,S12,S22,S32,XJ,RA,B,XKB,XKI,R,
     2ZX,ZY4,ZY5,ZY6,ZI,C1,C2,YR61
      COMMON/II/IP,JS,JC,KC,MC,NS,MPRINT,IC12
      DERY(1)=XK1*(Y61+YR61*X-Y(6))*S1
      DERY(2)=XK2(JS)*(A1*(1.DO-S12)+S12*(Y(1)-XKD1*Y(6))-Y(5))*S2
      DERY(3)=XK3(JS)*(A2*(1.DO-S22)+S22*(Y(2)-XKD2(JS)*Y(5))-Y(4))*S3
      DERY(4)=-ROL*Y(4)+(A3*(1.DO-S32)+S32*(Y(3)-XKD3(JS)*Y(4)))/XLA
     1-BKLA*Y(5)
      DERY(5)=-BOJ*Y(5)+XKOJ*Y(4)+(C1+C2*X)/XJ
      DERY(6)=Y(5)
      RETURN
      END
      SUBROUTINE OUTP(X,Y,DERY,IHLF,NDIM,PRMT)
      IMPLICIT REAL*8(A-H,O-Y)
      DIMENSION Y(6),DERY(6),PRMT(5),XK2(3),XKD2(3),XK3(3),XKD3(3),R(3),
     1ZX(600),ZY4(600),ZY5(600),ZY6(600),ZI(600)
      COMMON XK1,S1,A1,XKD1,XK2,S2,A2,XKD2,XK3,S3,A3,XKD3,ROL,XLA,BKLA,
     1BOJ,XKOJ,U1,U2,U3,XL1,XL2,XL3,Y61,S12,S22,S32,XJ,RA,B,XKB,XKI,R,
     2ZX,ZY4,ZY5,ZY6,ZI,C1,C2,YR61
      COMMON/II/IP,JS,JC,KC,MC,NS,MPRINT,IC12
      IF(IP.EQ.1)GO TO 50
```

```
      MC=MC+1
      TA1=XK1*(Y61+YR61*X-Y(6))
      T1=Y(1)-XKD1*Y(6)
      IF(T1.GT.U1)GO TO 109
      GO TO 101
  109 IF(IU1.EQ.1)GO TO 106
      A1=U1
      IU1=1
      IL1=0
  106 IF(S12.EQ.1.DO)CALL STOP(X,PRMT)
      S12=0.DO
      JS=2
      IR1=1
      IT=1
      IF(TA1.LT.0.DO)GO TO 103
      GO TO 102
  101 IF(T1.GT.XL1)GO TO 104
      IF(IL1.EQ.1)GO TO 108
      IU1=0
      IL1=1
      A1=XL1
  108 IF(S12.EQ.1.DO)CALL STOP(X,PRMT)
      S12=0.DO
      JS=2
      IR1=1
      IT=2
      IF(TA1.GT.0.DO)GO TO 103
  102 IF(S1.EQ.1.DO)CALL STOP(X,PRMT)
      S1=0.DO
      GO TO 150
  103 IF(S1.EQ.0.DO)CALL STOP(X,PRMT)
      S1=1.DO
      GO TO 150
  104 IU1=0
      IL1=0
      IR1=0
      IF(S12.EQ.0.DO)CALL STOP(X,PRMT)
      IF(S1.EQ.0.DO)CALL STOP(X,PRMT)
      S12=1.DO
      S1=1.DO
      JS=1
  150 TA2=XK2(JS)*(A1*(1.DO-S12)+S12*(Y(1)-XKD1*Y(6))-Y(5))
      T2=Y(2)-XKD2(JS)*Y(5)
      IF(T2.GT.U2)GO TO 209
      GO TO 201
  209 IF(IU2.EQ.1)GO TO 206
      A2=U2
      IU2=1
      IL2=0
  206 IF(S22.EQ.1.DO)CALL STOP(X,PRMT)
      S22=0.DO
      JS=3
      IR2=1
      IF(TA2.LT.0.DO)GO TO 203
      GO TO 202
  201 IF(T2.GT.XL2)GO TO 204
      IF(IL2.EQ.1)GO TO 208
      IU2=0
      IL2=1
      A2=XL2
  208 IF(S22.EQ.1.DO)CALL STOP(X,PRMT)
      S22=0.DO
      JS=3
      IR2=1
      IF(TA2.GT.0.DO)GO TO 203
  202 IF(S2.EQ.1.DO)CALL STOP(X,PRMT)
      S2=0.DO
      GO TO 250
  203 IF(S2.EQ.0.DO)CALL STOP(X,PRMT)
      S2=1.DO
      GO TO 250
```

```
  204 IU2=0
      IL2=0
      IR2=0
      IF(S22.EQ.0.D0)CALL STOP(X,PRMT)
      IF(S2.EQ.0.D0)CALL STOP(X,PRMT)
      S22=1.D0
      S2=1.D0
      JS=1
      IF(IR1.EQ.1)JS=2
  250 T3=Y(3)-XKD3(JS)*Y(4)
      TA3=XK3(JS)*(A2*(1.D0-S22)+S22*(Y(2)-XKD2(JS)*Y(5))-Y(4))
      IF(T3.LT.U3)GO TO 301
      IF(IU3.EQ.1)GO TO 306
      A3=U3
      IU3=1
      IL3=0
  306 IF(S32.EQ.1.D0)CALL STOP(X,PRMT)
      S32=0.D0
      IF(TA3.LT.0.D0)GO TO 303
      GO TO 302
  301 IF(T3.GT.XL3)GO TO 304
      IF(IL3.EQ.1)GO TO 308
      IU3=0
      IL3=1
      A3=XL3
  308 IF(S32.EQ.1.D0)CALL STOP(X,PRMT)
      S32=0.D0
      IF(TA3.GT.0.D0)GO TO 303
  302 IF(S3.EQ.1.D0)CALL STOP(X,PRMT)
      S3=0.D0
      GO TO 350
  303 IF(S3.EQ.0.D0)CALL STOP(X,PRMT)
      S3=1.D0
      GO TO 350
  304 IU3=0
      IL3=0
      IF(S32.EQ.0.D0)CALL STOP(X,PRMT)
      IF(S3.EQ.0.D0)CALL STOP(X,PRMT)
      S32=1.D0
      JS=3
      S3=1.D0
      IF(S22.EQ.0.D0)GO TO 350
      IF(S12.EQ.0.D0)GO TO 401
      JS=1
      GO TO 350
  401 JS=2
      IT=3
  350 IF(MPRINT.EQ.1)WRITE(6,100)X,Y(3),Y(4),Y(5),Y(6),JS,IHLF,MC
      JC=JC+1
      IF(JC.EQ.NS)GO TO 450
  100 FORMAT(5D12.4,2I3,I5)
      GO TO 51
  450 KC=KC+1
      IF(KC.GT.600)GO TO 52
      ZX(KC)=X
      ZY4(KC)=Y(4)
      ZY5(KC)=Y(5)
      ZY6(KC)=Y(6)
      ZI(KC)=Y61+YR61*X
      IF(IC12.EQ.1)ZI(KC)=(C1+C2*X)/XJ
      JC=0
C       WRITE(6,750)X,Y(4),Y(5),Y(6)
C       WRITE(6,750)ZX(KC),ZY4(KC),ZY5(KC),ZY6(KC),KC
  750 FORMAT(4E13.4,I5)
  751 FORMAT(4D13.4)
      RETURN
   50 IP=0
   51 RETURN
   52 PRMT(5)=5.D0
      RETURN
      END
```

```
      SUBROUTINE STOP(X,PRMT)
      IMPLICIT REAL*8(A-H,O-Y)
      DIMENSION PRMT(5),XKD2(3),XK2(3),XKD3(3),XK3(3),R(3),
     1ZX(600),ZY4(600),ZY5(600),ZY6(600),ZI(600)
      COMMON XK1,S1,A1,XKD1,XK2,S2,A2,XKD2,XK3,S3,A3,XKD3,ROL,XLA,BKLA,
     1BOJ,XKOJ,U1,U2,U3,XL1,XL2,XL3,Y61,S12,S22,S32,XJ,RA,B,XKB,XKI,R,
     2ZX,ZY4,ZY5,ZY6,ZI,C1,C2,YR61
      COMMON/II/IP,JS,JC,KC,MC,NS,MPRINT,IC12
      IP=1
      PRMT(5)=1.D0
      PRMT(1)=X
      RETURN
      END
```

CAPSTAN

```
      IMPLICIT REAL*8(A-H,O-Y)
      DIMENSION Y(7),DERY(7),PRMT(5),AUX(16,7),ZX(100),ZY4(100),
     1ZY7(100)
      COMMON XK1,XK2,XK3,XKD1,XKD2,XKD3,XJL,XJM,BL,XKE,XK,AD,BD,CD,
     1O1,O2,PI,ZX,ZY4,ZY7
      COMMON/II/IP,MC,NS,MPRINT,JC,KC
      EXTERNAL FCT,OUTP
      DO 453 I=1,100
      ZX(I)=0.
      ZY4(I)=0.
  453 ZY7(I)=0.
      MC=0
      KC=0
C234567
      PI=4.D0*DATAN(1.D0)
      IP=0
      WRITE(6,*)'REAL PART OF ROOT IS'
      READ(5,*)A
      WRITE(6,*)'REAL PART = ',A
      WRITE(6,*)'IMAGINARY PART OF ROOT IS'
      READ(5,*)B
      WRITE(6,*)'IMAGINARY PART = ',B
      WRITE(6,*)'SUPPLY THREE COEFFICIENTS OF DISTURBANCE'
      READ(5,*)AD,BD,CD
      WRITE(6,*)'DISTURBANCE COEFFICIENTS ARE',AD,BD,CD
      WRITE(6,*)'SUPPLY TWO INPUT COEFFICENTS'
      READ(5,*)O1,O2
      WRITE(6,*)'INPUT COEFFICIENTS ARE',O1,O2
      WRITE(6,*)' ENTER NP AND NS'
      READ(5,*)NP,NS
      WRITE(6,*)'NP=',NP,'NS=',NS
      WRITE(6,*)' ENTER MPRINT'
      READ(5,*)MPRINT
      WRITE(6,*)'MPRINT=',MPRINT
      O1=2.D0*O1*PI/60.D0
      O2=2.D0*O2*PI/60.D0
      XJM=.023D0
      XJL=7.24D0
      XKE=6.92D0
      BL=10.D0
      XK=2857.6D0
      A2=A*A
```

```
      C=A2+B*B
      C22=C*C
      C1=6. D0*A
      C2=3. D0*C+12. D0*A2
      C3=12. D0*A*C+8. D0*A2*A
      C4=3. D0*C22+12. D0*A2*C
      C5=6. D0*A*C22
      C6=C22*C
      XKD3=XKE+XJM*BL/XJL-C1*XJM
      XK3=XK*(1. D0+XJM/XJL)+BL*XKE/XJL-BL*XKD3/XJL-C2*XJM
      XKD2=XK*(XKE+BL)/(XJL*XK3)-BL/XJL-C3*XJM/XK3
      XK2=-XKD2*BL/XJL-C4*XJM/XK3
      XKD1=BL/XK+XJM*XJL*C5/(XK2*XK3*XK)
      XK1=XJM*XJL*C6/(XK*XK2*XK3)
      WRITE(6,*)' K1= ',XK1
      WRITE(6,*)' K2= ',XK2
      WRITE(6,*)' K3= ',XK3
      WRITE(6,*)' KD1= ',XKD1
      WRITE(6,*)' KD2= ',XKD2
      WRITE(6,*)' KD3= ',XKD3
      WRITE(6,*)' OLD VALUE OF XJL = ',XJL
      WRITE(6,*)' ENTER NEW OR SAME XJL'
      READ(5,*)XJL
      WRITE(6,*)' XJL=',XJL
      NDIM=7
      DIM=NDIM
      ODIM=1. D0/DIM
      DO 5 I=1,NDIM
    5 DERY(I)=ODIM
      Y(6)=200. D0*PI/60. D0
      Y(4)=Y(6)
      Y(5)=0. D0
      Y(7)=-Y(6)*BL/XK
      Y(3)=XKE*Y(4)-XK*Y(7)
      Y(2)=XKD2*Y(7)
      Y(1)=XKD1*Y(6)+Y(7)
      PRMT(1)=0. D0
      PRMT(2)=4. D0
      PRMT(3)=. 001D0
      PRMT(4)=1. D-6
      CALL DHPCG(PRMT,Y,DERY,NDIM,IHLF,FCT,OUTP,AUX)
      CALL INITT(960)
      CALL TERM(2,1024)
      CALL CZAXIS(1)
      CALL BINITT
      CALL SLIMX(225,975)
      CALL NPTS(NP)
      CALL CHECK(ZX,ZY7)
      CALL DSPLAY(ZX,ZY7)
      CALL DINITY
      CALL YLOC(-100)
      CALL LINE(3)
      CALL CHECK(ZX,ZY4)
      CALL DSPLAY(ZX,ZY4)
      CALL TINPUT(I)
      CALL FINITT(0,700)
      STOP
      END
      SUBROUTINE FCT(X,Y,DERY)
      IMPLICIT REAL*8(A-H,O-Y)
      DIMENSION Y(7),DERY(7),ZX(100),ZY4(100),ZY7(100)
     COMMON XK1,XK2,XK3,XKD1,XKD2,XKD3,XJL,XJM,BL,XKE,XK,AD,BD,CD,
    1O1,O2,PI,ZX,ZY4,ZY7
     COMMON/II/IP,MC,NS,MPRINT,JC,KC
      OM=O1+O2*X
      D=AD+BD*X+CD*X*X
      DERY(1)=XK1*(OM-Y(6))
      DERY(2)=XK2*(Y(1)-XKD1*Y(6)-Y(7))
      DERY(3)=XK3*(Y(2)-XKD2*Y(7)-Y(5))
      DERY(4)=(Y(3)-XKD3*Y(5)-XKE*Y(4)+XK*Y(7))/XJM
      DERY(5)=(D-BL*Y(6)-XK*Y(7))/XJL-(Y(3)-XKD3*Y(5)-XKE*Y(4)
    1+XK*Y(7))/XJM
```

```
      DERY(6)=(D-BL*Y(6)-XK*Y(7))/XJL
      DERY(7)=Y(5)
      RETURN
      END
      SUBROUTINE OUTP(X,Y,DERY,IHLF,NDIM,PRMT)
      IMPLICIT REAL*8(A-H,O-Y)
      DIMENSION Y(7),DERY(7),PRMT(5),ZX(100),ZY4(100),ZY7(100)
     COMMON XK1,XK2,XK3,XKD1,XKD2,XKD3,XJL,XJM,BL,XKE,XK,AD,BD,CD,
    1O1,O2,PI,ZX,ZY4,ZY7
     COMMON/II/IP,MC,NS,MPRINT,JC,KC
      MC=MC+1
      Y6=Y(6)*60.DO/(2.DO*PI)
      Y4=Y(4)*60.DO/(2.DO*PI)
      F=Y(3)-XKD3*Y(5)
  100 FORMAT(' ',5D13.4,I4,I6)
      IF(MPRINT.EQ.1)WRITE(6,100)X,Y4,Y(5),Y6,Y(7),IHLF,MC
      JC=JC+1
      IF(JC.EQ.NS)GO TO 450
      RETURN
  450 KC=KC+1
      IF(KC.GT.100)RETURN
      ZX(KC)=X
      ZY4(KC)=Y6
      ZY7(KC)=Y(7)
      JC=0
      IP=0
      RETURN
      END
```

DIG

```
C MAIN PROGRAM FOR DIGITAL SIMULATION
C SPRING - MASS SYSTEM WITH IESF CONTROL
      IMPLICIT REAL*8(A-H,O-Y)
      DIMENSION X(2),DERX(2),PRMT(5),AUX(8,2),B(3,3),COM(2,3),PHI(2,2),
     1THETA(2),ZT(500),ZY42(500),ZF2(500),ZR2(500),ZDIS(500),Y(4),
     2DERY(4),PRM1(5),AUX3(8,4)
      COMMON OMN,DZETA,OMN2,Y12,Y22,Y32,Y42,F2,R,T,TIME,R2,ZT,ZY42,ZF2,
     1ZR2,Y42M,R2M,TPI,DA,DB,DF,DC,DIS,ZDIS,Y52
      COMMON/II/IP,MC,NS,JC,KC,MPRINT,NP,IDST,IINT
      COMMON/XK/XK1,XK2,XKD1,XKD2
      EQUIVALENCE(PHI(1),COM(1)),(THETA(1),COM(5))
      EXTERNAL FCT,OUTP,FCT2,OUTP2,FCT3,OUTP3
C COMPUTE TRANSITION MATRIX OF CONTINUOUS PLANT
C TRANSITION MATRIX IS FOR SAMPLING INTERVAL T
      WRITE(6,*)'ENTER SAMPLING INTERVAL T'
      Y42M=0.DO
      R2M=0.DO
      JC=0
      READ(5,*)T
      TO =T
      WRITE(6,*)'SAMPLING INTERVAL IS',T,'NORMALIZED UNITS'
      PRMT(2)=T
      WRITE(6,*)'ENTER COMPUTING INTERVAL FOR DKGS'
      READ(5,*)PRMT(3)
      WRITE(6,*)'COMPUTING INTERVAL IS',PRMT(3),'NORMALIZED UNITS'
      WRITE(6,*)'ENTER NATURAL FREQUENCY,1 OR 0'
      READ(5,*)OMN1
      OMNO=OMN1
      WRITE(6,*)'NATURAL FREQUENCY IS',OMN1,'NORMALIZED HZ'
      PI=4.DO*DATAN(1.DO)
      TPI =2.DO*PI
      OMN=TPI*OMN1
      OMN2=OMN*OMN
```

```
      WRITE(6,*)'ENTER DAMPING RATIO; IF OMN=0, THEN RATIO IS C/M'
      READ(5,*)DZETA
      DZETAO=DZETA
      WRITE(6,*)'ZETA=',DZETA
      PRMT(4)=1.D-8
      NDIM=2
      DO 15 I=1,3
      DO 15 J=1,3
   15 B(I,J)=0.DO
      DO 16 I=1,3
   16 B(I,I)=1.DO
      DO 17 I=1,3
      X(1)=B(1,I)
      X(2)=B(2,I)
      DERX(1)=.5DO
      DERX(2)=.5DO
      F2=B(3,I)
      PRMT(1)=0.DO
      CALL DKGS(PRMT,X,DERX,NDIM,IHLF,FCT,OUTP,AUX)
      COM(1,I)=X(1)
   17 COM(2,I)=X(2)
      TH1=THETA(1)
      TH2=THETA(2)
      WRITE(6,*)TH1,TH2
      P11=PHI(1,1)
      P12=PHI(1,2)
      P21=PHI(2,1)
      P22=PHI(2,2)
      WRITE(6,100)P11,P12,P21,P22
C PHI AND THETA HAVE BEEN COMPUTED
C CHOOSE A DISCRETE APPROACH OR A CONTINUOUS APPROXIMATION
      WRITE(6,*)'IF DISCRETE APPROACH,ENTER 1; IF CONTINUOUS
     1 APPROXIMATION, ENTER 0'
      READ(5,*)IDIS
      IF(IDIS.EQ.1)GO TO 21
      WRITE(6,*)'YOU HAVE PICKED THE CONTINOUS APPROXIMATION'
      WRITE(6,*)'THIS PROGRAM HANDLES REPEATED ROOTS ON REAL AXIS'
      WRITE(6,*)'ENTER ABSOLUTE VALUE OF ROOT'
      READ(5,*)A
      WRITE(6,*)'A=',A
      XKD2=4.DO*A-2.DO*DZETA*OMN
      IF(OMN.EQ.0.DO)XKD2=4.DO*A-2.DO*DZETA
      XK2=6.DO*A*A-OMN*OMN
      XKD1=4.DO*A*A*A/XK2
      XK1=A*A*A*A/XK2
      EA=DEXP(-A*T)
      EA2=EA*EA
      EA3=EA2*EA
      EA4=EA2*EA2
      PRM1(1)=0.DO
      PRM1(2)=100.DO
      PRM1(3)=.01DO
      PRM1(4)=1.D-6
      DO 888 I=1,4
      Y(I)=0.DO
  888 DERY(I)=.25DO
      NDIT=4
C       CALL DKGS(PRM1,Y,DERY,NDIT,IHLF,FCT3,OUTP3,AUX3)
      GO TO 20
   21 WRITE(6,*)'YOU HAVE PICKED THE DISCRETE APPROACH'
      WRITE(6,*)'IF YOU WANT VERY FAST RESPONSE,ENTER 1;
     1 IF SLOWER, ENTER 0'
      READ(5,*)IFAST
      WRITE(6,*)'IFAST=',IFAST
      IF(IFAST.EQ.1)GO TO 25
      WRITE(6,*)'ENTER REAL ROOT'
      READ(5,*)EA
      WRITE(6,*)'REAL ROOT IS',EA
      A=DLOG(EA)/T
      WRITE(6,*)' S PLANE ROOT IS',A
      EA2=EA*EA
```

```
      EA3=EA2*EA
      EA4=EA2*EA2
      EA5=EA3*EA2
      GO TO 26
   25 EA=0.DO
   26 WRITE(6,*)'IF YOU HAVE COMPLETE STATE FEEDBACK, ENTER 1,
     1 ELSE ENTER O'
      READ(5,*)ICSF
      IF(ICSF.NE.1)GO TO 171
      XKD2=(P22+P11+2.DO-4.DO*EA)/TH1
      XK1N=(1.DO+EA4-4.DO*EA3+6.DO*EA2-4.DO*EA)/(T*T*(P21*TH1
     1-P11*TH2+TH2))
      A11=T*(P21*TH1-TH2*P11-TH2)
      A21=T*TH2
      A12=T*(-P22*TH1+TH2*P12-TH1)
      A22=T*TH1
      B1=-XKD2*(2.DO*P22*TH1-2.DO*TH2*P12+TH1)+2.DO*P22*P11+P22
     1-2.DO*P21*P12+P11-XK1N*T*T*TH2-4.DO*EA3
      B2=-XKD2*(-P22*TH1+TH2*P12-2.DO*TH1)-P22*P11-2.DO*P22+P21*P12
     1-2.DO*P11-1.DO+6.DO*EA2
      DET=A11*A22-A12*A21
      XK2=(A11*B2-B1*A21)/DET
      XKD1=(B1*A22-A12*B2)/(DET*XK2)
      XK1=XK1N/XK2
      GO TO 420
   20 ICSF=1
  420 WRITE(6,*)'HERE ARE K1,K2,KD1,KD2'
      WRITE(6,898)XK1,XK2,XKD1,XKD2
  898 FORMAT(/'  K1 = ',1PE11.4,'      K2 = ',1PE11.4//' KD1 = ',
     11PE11.4,'    KD2 = ',1PE11.4/)
      GO=XK1*XK2*T*T*(P21*TH1-TH2*P11)+XKD1*XK2*T*(-P21*TH1+TH2*P11)
     1+XK2*T*(P22*TH1-TH2*P12)+XKD2*(-P22*TH1+TH2*P12)+P22*P11
     2-P21*P12-EA4
      G1=XK1*XK2*T*T*TH2+XKD1*XK2*T*(P21*TH1-TH2*P11-TH2)+XK2*T*
     1(-P22*TH1+TH2*P12-TH1)+XKD2*(2.DO*P22*TH1-2.DO*TH2*P12+TH1)
     2-2.DO*P22*P11-P22+2.DO*P21*P12-P11+4.DO*EA3
      G2=XKD1*XK2*T*TH2+XK2*T*TH1+XKD2*(-P22*TH1+TH2*P12-2.DO*TH1)
     1+P22*P11+2.DO*P22-P21*P12+2.DO*P11+1.DO-6.DO*EA2
      G3=XKD2*TH1-P22-P11-2.DO+4.DO*EA
      WRITE(6,100)GO,G1,G2,G3
      GO TO 172
171   WRITE(6,*)'FOR TWO STAGES OF INTEGRAL COMPENSATION, ENTER 2,
     1ELSE 1'
      READ(5,*)IINT
      IF(IINT.EQ.2)GO TO 671
      WRITE(6,*)'YOU HAVE 1 STAGE OF INTEGRAL COMPENSATION'
      XK1=(EA4-4.DO*EA3+6.DO*EA2-4.DO*EA+1.DO)/
     1(T*T*(P21*TH1-TH2*P11+TH2))
      XK4=(XK1*T*T*TH2*TH2+4.DO*EA3*TH2+6.DO*EA2*(P21*TH1-TH2*P11
     1-TH2)+4.DO*EA*(-P22*P21*TH1+P21*P12*TH2-P21*P11*TH1-P21*TH1
     2+TH2*P11*P11+TH2*P11+TH2)+P22*(P22*P21*TH1-P21*TH2*P12+P21*P11
     3*TH1+P21*TH1)+P21*(P21*P12*TH1-2.DO*TH2*P12*P11-TH2*P12+P11*
     4P11*TH1+P11*TH1+TH1)+P11*TH2*(-P11*P11-P11-1.DO)-TH2)/(P21*(P22*
     5TH2*TH1+P21*TH1*TH1-TH2*TH2*P12-TH2*P11*TH1))
      XK3=(-XK4*TH2-4.DO*EA+P22+P11+2.DO)/T
      XK2=(XK4*XK4*TH2*TH2+XK4*(4.DO*EA*TH2-2.DO*P22*TH2-P21*TH1
     1-TH2*P11-TH2)+6.DO*EA*EA-4.DO*EA*P22-4.DO*EA*P11-4.DO*EA+
     2P22*P22+P22*P11+P22+P12*P21+P11*P11+P11+1.DO)/(T*TH2)
      WRITE(6,899)XK1,XK2,XK3,XK4
  899 FORMAT(/' K1 = ',1PE11.4,'     K2 = ',1PE11.4//' K3 = ',1PE11.4,
     1'     K4 = ',1PE11.4/)
      TO=XK1*T*T*(P21*TH1-TH2*P11)+XK2*T*(-P21*TH1+TH2*P11)+XK3*XK4*T*
     1(-P21*TH1+TH2*P11)+XK3*T*(-P22*P11+P21*P12)+XK4*(P21*TH1-TH2*P11
     2)-EA4+P22*P11-P21*P12
      T1=XK1*T*T*TH2+XK2*T*(P21*TH1-TH2*P11-TH2)+XK3*XK4*T*(P21*TH1
     1-TH2*P11-TH2)+XK3*T*(P22*P11+P22-P21*P12+P11)+XK4*(-2.DO*P21*
     2TH1+2.DO*TH2*P11+TH2)+4.DO*EA3-2.DO*P22*P11-P22+2.DO*P21*P12-P11
      T2=XK2*T*TH2+XK3*XK4*T*TH2-(XK3*T)*(P22+P11+1.DO)+XK4*(P21*TH1
     1-TH2*P11-2.DO*TH2)-6.DO*EA2+P22*P11+2.DO*P22-P12*P21+2.DO*P11
     2+1.DO
      T3=XK3*T+XK4*TH2+4.DO*EA-P22-P11-2.DO
```

```
      WRITE(6,*)'TO= ',TO,'T1= ',T1,'T2= ',T2,'T3= ',T3
      GO TO 172
671   RV=EA-1
      P1M=1.DO-P11
      P2M=1.DO-P22
      S12=P2M+P1M
      RV2=RV*RV
      RV3=RV2*RV
      RV4=RV2*RV2
      RV5=RV2*RV3
      P212=P21*P12
      P1M2=P1M*P1M
      P1M3=P1M*P1M2
      P1M4=P1M2*P1M2
      P1M5=P1M2*P1M3
      TH22=TH2*TH2
      TH23=TH2*TH22
      TH24=TH22*TH22
      TH12=TH1*TH1
      TH13=TH1*TH12
      TH14=TH12*TH12
      XK1=(-RV5)/(T**3*(P21*TH1+TH2*P1M))
      XL1=-TH23*S12*P212*P1M2+TH23*(-P212*P1M3-5.DO*P212*P1M2*RV-P1M5
     1-5.DO*P1M4*RV-10.DO*P1M3*RV2-10.DO*P1M2*RV3-
     2 5.DO*P1M*RV4-RV5)
      XL2=-TH22*TH1*S12*S12*P21*P1M2+TH22*TH1*S12*P21*P1M*(-2.DO*P212+
     1P1M2-5.DO*P1M*RV)
      XL3=TH22*TH1*P21*(-3.DO*P212*P1M2-10.DO*P212*P1M*RV-3.DO*P1M4
     1-10.DO*P1M3*RV-30.DO*P1M2*RV2-20.DO*P1M*RV3-5.DO*RV4)
      XL4=-2.DO*TH2*TH12*S12*S12*P21*P21*P1M+TH2*TH12*S12*P21*P21*
     1(-P212+2.DO*P1M2-10.DO*P1M*RV)
      XL5=TH2*TH12*P21*P21*(-3.DO*P212*P1M-5.DO*P212*RV-3.DO*P1M3-5.DO
     1*P1M2*RV-30.DO*P1M*RV2-10.DO*RV3)
     2-TH13*S12*S12*P21**3
      XL6=TH13*S12*P21*P21*P21*(P1M-5.DO*RV)+TH13*P21*P21*P21*(-P212-
     1P1M2-10.DO*RV2)
      XL7=TH24*P212*P1M*P1M+TH23*TH1*S12*P21*P1M2+2.DO*TH23*TH1*P21*
     1P1M*(P212-P1M2)+2.DO*TH22*TH12*S12*P21*P21*P1M+TH22*TH12*P21*
     2P21*(P212-5.DO*P1M2)+TH2*TH13*S12*P21*P21*P21-4.DO*TH2*TH13*P21
     3*P21*P21*P1M-TH14*P21**4
      XK5=(XL1+XL2+XL3+XL4+XL5+XL6)/XL7
      XK4=(XK5*XK5*TH22+2.DO*XK5*TH2*S12+XK5*TH2*(-P1M+5.DO*RV)-
     1XK5*TH1*P21+S12*S12+S12*(-P1M+5.DO*RV)+P212+P1M2+10.DO*RV2
     2)/(TH2*T)
      XK1=(-RV5)/(T*T*T*(P21*TH1+TH2*P1M))
      XK3=(-XK5*TH2-S12-5.DO*RV)/T
      XK2=(-XK1*T*T*T*TH2-XK3*XK4*T*T*P21*TH1-XK3*XK4*T*T*TH2*P1M
     1+5.DO*RV4)/(T*T*(P21*TH1+TH2*P1M))
      WRITE(6,399)XK1,XK2,XK3,XK4,XK5
  399 FORMAT(/' K1 = ',1PE11.4,'     K2 = ',1PE11.4,'     K3 = ',
     11PE11.4//' K4 = ',1PE11.4,'     K5 = ',1PE11.4/)
      T4=XK3*T+XK5*TH2+S12+5.DO*RV
      T3=XK3*XK5*T*TH2+XK3*T*S12+XK4*T*TH2+XK5*P21*TH1+XK5*TH2*P1M
     1-P212+S12*P1M-P1M2-10.DO*RV2
      T2=(XK2*T*TH2+XK3*XK4*T*TH2+XK3*XK5*P21*TH1+XK3*XK5*TH2*P1M-XK3*
     1P212+XK3*S12*P1M-XK3*P1M2+XK4*P21*TH1+XK4*TH2*P1M)*T+10.DO*RV3
      T1=(XK1*T*TH2+XK2*P21*TH1+XK2*TH2*P1M+XK3*XK4*P21*TH1+XK3*XK4*TH2
     1*P1M)*T*T-5.DO*RV4
      TO=XK1*T**3*(P21*TH1+TH2*P1M)+RV5
      WRITE(6,*)'TO= ',TO,'T1= ',T1,'T2= ',T2,'T3= ',T3,'T4= ',T4
  172 WRITE(6,*)' C1 IS SIZE OF STEP, C2 IS SLOPE OF RAMP,C3 IS
     1 AMPLITUDE OF SINE INPUT, C4 IS FREQUENCY IN HZ'
      WRITE(6,*)'SUPPLY FOUR INPUT COEFFICIENTS'
      READ(5,*)C1,C2,C3,C4
      WRITE(6,*)'C1=',C1,'C2=',C2,'C3=',C3,'C4=',C4
      WRITE(6,*)'IF YOU WANT TO CHANGE NATURAL FREQUENCY,ENTER 1;
     1 IF NOT,ENTER 0'
      READ(5,*)IWN
      IF(IWN.EQ.1)GO TO 31
      GO TO 32
   31 WRITE(6,*)'ENTER ADDITION TO OMEGA NAT'
      READ(5,*)ADOMN1
```

```
      WRITE(6,*)'ADOMN=',ADOMN1
      OMN1=OMN1+ADOMN1
      OMN=TPI*OMN1
      OMN2=OMN*OMN
   32 WRITE(6,*)'IF YOU WANT TO CHANGE DAMPING RATIO, ENTER 1; IF
     1 NOT, ENTER 0'
      READ(5,*)IZETA
      IF(IZETA.EQ.1)GO TO 33
      IF(IWN.EQ.0)GO TO 34
      GO TO 37
   33 WRITE(6,*)'ENTER ADDITION TO ZETA'
      READ(5,*)ADZET
      WRITE(6,*)'ADDITION TO ZETA IS',ADZET
      DZETA=DZETA+ADZET
   37 WRITE(6,*)'OMN=',OMN,'ZETA=',DZETA
  100 FORMAT(4D16.7)
   34 TIME=0.D0
      WRITE(6,*)' TO MODIFY THE SAMPLING INTERVAL, ENTER 1; ELSE 0'
      READ(5,*)ITMOD
      IF(ITMOD.EQ.0)GO TO 79
      IF(ITMOD.EQ.1)WRITE(6,*)' ENTER ADDITION TO T'
      READ(5,*)TMOD
      T=T+TMOD
      WRITE(6,*)'NEW SAMPLING INTERVAL IS  ',T
   79 IF(IWN.EQ.1.OR.IZETA.EQ.1.OR.ITMOD.EQ.1)GO TO 77
      GO TO 78
   77 DO 67 I=1,3
      X(1)=B(1,I)
      X(2)=B(2,I)
      DERX(1)=.5D0
      DERX(2)=.5D0
      F2=B(3,I)
      PRMT(1)=0.D0
      CALL DKGS(PRMT,X,DERX,NDIM,IHLF,FCT,OUTP,AUX)
      COM(1,I)=X(1)
   67 COM(2,I)=X(2)
      TH1=THETA(1)
      TH2=THETA(2)
      P11=PHI(1,1)
      P12=PHI(1,2)
      P21=PHI(2,1)
      P22=PHI(2,2)
      WRITE(6,100)TH1,TH2
      WRITE(6,100)P11,P12,P21,P22
   78 WRITE(6,*)'IF THERE IS A DISTURBANCE, ENTER 1, ELSE ENTER 0'
      READ(5,*)IDST
      IF(IDST.EQ.1)GO TO 737
      GO TO 738
  737 WRITE(6,*)'ENTER FOUR FORCE DISTURBANCE COEFFICIENTS, STEP,
     1 RAMP, AMPLITUDE, FREQUENCY'
      READ(5,*)DA,DB,DC,DF
      WRITE(6,*)DA,DB,DC,DF
  738 WRITE(6,*)'ENTER NP AND NS'
      READ(5,*)NP,NS
      WRITE(6,*)'NP=',NP,'NS=',NS
      WRITE(6,*)'FOR CONTINOUS INTEGRATION, ENTER 1, ELSE ENTER 0'
      READ(5,*)ICT
      MC=0
      KC=0
      WRITE(6,*)' CLEAR SCREEN'
      READ(6,*)IFAC
      WRITE(6,600)OMNO
  600 FORMAT(/' UNDAMPED NATURAL FREQUENCY = ',1PE11.4)
      IF(IWN.EQ.1)WRITE(6,601)OMN1
  601 FORMAT(/' NATURAL FREQUENCY CHANGED TO ',1PE11.4)
      WRITE(6,602)DZETAO
  602 FORMAT(/' DAMPING RATIO = ',1PE11.4)
      IF(IZETA.EQ.1)WRITE(6,603)DZETA
  603 FORMAT(/' DAMPING RATIO CHANGED TO ',1PE11.4)
      WRITE(6,604)TO
  604 FORMAT(/' SAMPLING INTERVAL = ',1PE11.4)
      IF(ITMOD.EQ.1)WRITE(6,605)T
```

```
605 FORMAT(/' SAMPLING INTERVAL CHANGED TO ',1PE11.4)
    IF(IDST.EQ.1)WRITE(6,606)DA,DB,DC,DF
606 FORMAT(/' DISTURBANCE COEFFICIENTS'/4(4X,1PE11.4))
    WRITE(6,607)EA
    WRITE(6,608)A
607 FORMAT(/' EA = ',1PE11.4)
608 FORMAT(/' A = ',1PE11.4)
    IF(IDIS.EQ.1.AND.IINT.EQ.1)WRITE(6,899)XK1,XK2,XK3,XK4
    IF(IDIS.EQ.1.AND.IINT.EQ.2)WRITE(6,399)XK1,XK2,XK3,XK4,XK5
    IF(IDIS.EQ.0)WRITE(6,898)XK1,XK2,XKD1,XKD2
    IF(IDIS.EQ.1.AND.ICSF.EQ.1)WRITE(6,898)XK1,XK2,XKD1,XKD2
    IF(IINT.EQ.2)WRITE(6,*)'INCOMPLETE STATE FEEDBACK - 2 STAGES OF
   1 INTEGRAL COMPENSATION'
    IF(IINT.EQ.1)WRITE(6,*)'INCOMPLETE STATE FEEDBACK - 1 STAGE OF
   1 INTEGRAL COMPENSATION'
    IF(IDIS.EQ.0)WRITE(6,*)' CONTINUOUS APPROXIMATION'
    IF(IDIS.EQ.1.AND.ICSF.EQ.1)WRITE(6,*)' COMPLETE STATE FEEDBACK'
    WRITE(6,625)
625 FORMAT(//' DISPLACEMENT -  INPUT AND RESPONSE'////' INPUT
   1FORCE     TIME     RESPONSE     FORCE     COPY'//)
    READ(6,*)IFAC
    WRITE(6,*)'ENTER MPRINT'
    READ(5,*)MPRINT
    WRITE(6,*)'MPRINT=',MPRINT
    Y12=0.D0
    Y22=0.D0
    Y32=0.D0
    Y42=0.D0
    Y11=0.D0
    Y21=0.D0
    Y31=0.D0
    Y41=0.D0
    Y51=0.D0
    Y52=0.D0
    R2=C1+C2*TIME+C3*DSIN(C4*TPI*TIME)
    TIME=TIME+T
    F2=Y21-XKD2*Y31
    ICON=1
    TIME=0.D0
    CALL OUTP1
    IF(IDST.EQ.1)GO TO 112
    IF(ICT.EQ.1)GO TO 112
    DO 10 I=1,10000
    XI=I
    R=C1+C2*TIME+C3*DSIN(C4*TPI*TIME)
    TIME=TIME+T
    R2=C1+C2*TIME+C3*DSIN(C4*TPI*TIME)
C INTEGRATION IN CONTROL ALGORITHM
    IF(ICSF.EQ.0)GO TO 173
    Y12=Y11+XK1*(R-Y41)*T
    Y22=Y21+XK2*(Y11-XKD1*Y41-Y31)*T
    F2=Y21-XKD2*Y31
    GO TO 174
C INCOMPLETE STATE FEEDBACK - 1 STAGE INTEGRAL COMPENSAION
173 IF(IINT.EQ.2)GO TO 273
    Y12=Y11+XK1*T*(R-Y41)
    Y22=Y21*(1.D0-XK3*T)+T*(Y11-XK2*Y41)
    F2=Y21-XK4*Y41
    GO TO 174
C INCOMPLETE STATE FEEDBACK - 2 STAGSE INTEGRAL COMPENSAION
273 Y12=Y11+XK1*(R-Y51)*T
    Y22=Y21*(1.D0-XK3*T)+T*(Y11-XK2*Y51)
    Y32=Y31+T*(Y21-XK4*Y51)
    F2=Y31-XK5*Y51
    Y42=P11*Y41+P12*Y51+TH1*F2
    Y52=P21*Y41+P22*Y51+TH2*F2
    GO TO 181
C INTEGRATE EQUATIONS OF MOTION USING TRANSITION MATRIX
174 Y32=P11*Y31+P12*Y41+TH1*F2
    Y42=P21*Y31+P22*Y41+TH2*F2
181 CALL OUTP1
    IF(KC.GT.NP)GO TO 11
    Y11=Y12
```

```
      Y21=Y22
      Y31=Y32
      Y51=Y52
   10 Y41=Y42
      GO TO 11
  112 DO 110 I=1,10000
      XI=I
      R=C1+C2*TIME+C3*DSIN(C4*TPI*TIME)
      TIME=TIME+T
      R2=C1+C2*TIME+C3*DSIN(C4*TPI*TIME)
C INTEGRATION IN CONTROL ALGORITHM
      IF(ICSF.EQ.0)GO TO 175
      Y12=Y11+XK1*(R-Y41)*T
      Y22=Y21+XK2*(Y11-XKD1*Y41-Y31)*T
      F2=Y21-XKD2*Y31
      X(1)=Y31
      X(2)=Y41
      GO TO 176
C INCOMPLETE STATE FEEDBACK - 1 STAGE INTEGRAL COMPENSAION
  175 IF(IINT.EQ.2)GO TO 275
      Y12=Y11+XK1*T*(R-Y41)
      Y22=Y21*(1.DO-XK3*T)+T*(Y11-XK2*Y41)
      F2=Y21-XK4*Y41
      X(1)=Y31
      X(2)=Y41
      GO TO 176
C INCOMPLETE STATE FEEDBACK - 2 STAGSE INTEGRAL COMPENSAION
  275 Y12=Y11+XK1*(R-Y51)*T
      Y22=Y21*(1.DO-XK3*T)+T*(Y11-XK2*Y51)
      Y32=Y31+T*(Y21-XK4*Y51)
      F2=Y31-XK5*Y51
      X(1)=Y41
      X(2)=Y51
C INTEGRATE EQUATIONS OF MOTION USING RUNGE-KUTTA
  176 PRMT(1)=TIME-T
      PRMT(2)=TIME
      DERX(1)=.5DO
      DERX(2)=.5DO
      CALL DKGS(PRMT,X,DERX,NDIM,IHLF,FCT2,OUTP2,AUX)
      IF(IINT.EQ.2)GO TO 177
      Y32=X(1)
      Y42=X(2)
      GO TO 178
  177 Y42=X(1)
      Y52=X(2)
  178 IF(KC.GT.NP)GO TO 11
      Y11=Y12
      Y21=Y22
      Y31=Y32
      Y51=Y52
  110 Y41=Y42
   11 CALL INITT(960)
      CALL TERM(2,1024)
      CALL CZAXIS(1)
      CALL BINITT
      CALL SLIMX(300,975)
      CALL NPTS(NP)
      CALL LINE(-11)
      CALL XFRM(2)
      CALL YFRM(2)
      CALL CHECK(ZT,ZF2)
      CALL DSPLAY(ZT,ZF2)
      IF(IDST.EQ.0)GO TO 751
      CALL DINITY
      CALL DINITX
      CALL LINE(2)
      CALL XFRM(2)
      CALL YFRM(2)
      CALL CPLOT(ZT,ZDIS)
  751 CALL DINITY
      CALL DINITX
      CALL YLOC(-150)
      CALL LINE(3)
```

```
      CALL XFRM(2)
      CALL YFRM(2)
      CALL CHECK(ZT,ZY42)
      CALL DSPLAY(ZT,ZY42)
      CALL DINITY
      CALL DINITX
      CALL LINE(2)
      CALL XFRM(2)
      CALL YFRM(2)
      CALL CPLOT(ZT,ZR2)
  752 CALL TINPUT(IV)
      CALL FINITT(0,700)
      STOP
      END
      SUBROUTINE FCT(TAU,X,DERX)
      IMPLICIT REAL*8(A-H,O-Y)
      DIMENSION X(2),DERX(2),ZT(500),ZY42(500),ZF2(500),ZR2(500),ZDIS
     1(500)
      COMMON OMN,DZETA,OMN2,Y12,Y22,Y32,Y42,F2,R,T,TIME,R2,ZT,ZY42,ZF2,
     1ZR2,Y42M,R2M,TPI,DA,DB,DF,DC,DIS,ZDIS,Y52
      COMMON/II/IP,MC,NS,JC,KC,MPRINT,NP,IDST,IINT
      DERX(1)=F2-2.DO*DZETA*OMN*X(1)-OMN2*X(2)
      IF(OMN.EQ.0.DO)DERX(1)=F2-2.DO*DZETA*X(1)
      DERX(2)=X(1)
      RETURN
      END
      SUBROUTINE FCT3(TAU,Y,DERY)
      IMPLICIT REAL*8(A-H,O-Y)
      DIMENSION Y(4),DERY(4),ZT(500),ZY42(500),ZF2(500),ZR2(500),ZDIS
     1(500)
      COMMON OMN,DZETA,OMN2,Y12,Y22,Y32,Y42,F2,R,T,TIME,R2,ZT,ZY42,ZF2,
     1ZR2,Y42M,R2M,TPI,DA,DB,DF,DC,DIS,ZDIS,Y52
      COMMON/II/IP,MC,NS,JC,KC,MPRINT,NP,IDST,IINT
      COMMON/XK/XK1,XK2,XKD1,XKD2
      DERY(1)=XK1*(1.DO-Y(4))
      DERY(2)=XK2*(Y(1)-XKD1*Y(4)-Y(3))
      DERY(3)=-2.DO*DZETA*OMN*Y(3)-OMN2*Y(4)+Y(2)-XKD2*Y(3)
      DERY(4)=Y(3)
      RETURN
      END
      SUBROUTINE FCT2(TAU,X,DERX)
      IMPLICIT REAL*8(A-H,O-Y)
      DIMENSION X(2),DERX(2),ZT(500),ZY42(500),ZF2(500),ZR2(500),ZDIS
     1(500)
      COMMON OMN,DZETA,OMN2,Y12,Y22,Y32,Y42,F2,R,T,TIME,R2,ZT,ZY42,ZF2,
     1ZR2,Y42M,R2M,TPI,DA,DB,DF,DC,DIS,ZDIS,Y52
      COMMON/II/IP,MC,NS,JC,KC,MPRINT,NP,IDST,IINT
      DIS=DA+DB*TAU+DC*DSIN(DF*TPI*TAU)
      DERX(1)=F2-2.DO*DZETA*OMN*X(1)-OMN2*X(2)+DIS
      IF(OMN.EQ.0.DO)DERX(1)=F2-2.DO*DZETA*X(1)
      DERX(2)=X(1)
      RETURN
      END
      SUBROUTINE OUTP(TAU,X,DERX,IHLF,NDIM,PRMT)
      IMPLICIT REAL*8(A-H,O-Y)
      DIMENSION X(2),DERX(2),ZT(500),ZY42(500),ZF2(500),ZR2(500),ZDIS
     1(500),PRMT(5)
      COMMON OMN,DZETA,OMN2,Y12,Y22,Y32,Y42,F2,R,T,TIME,R2,ZT,ZY42,ZF2,
     1ZR2,Y42M,R2M,TPI,DA,DB,DF,DC,DIS,ZDIS,Y52
      COMMON/II/IP,MC,NS,JC,KC,MPRINT,NP,IDST,IINT
      RETURN
      END
      SUBROUTINE OUTP3(TAU,Y,DERY,IHLF,NDIT,PRM1)
      IMPLICIT REAL*8(A-H,O-Y)
      DIMENSION Y(4),DERY(4),ZT(500),ZY42(500),ZF2(500),ZR2(500),ZDIS
     1(500),PRM1(5)
      COMMON OMN,DZETA,OMN2,Y12,Y22,Y32,Y42,F2,R,T,TIME,R2,ZT,ZY42,ZF2,
     1ZR2,Y42M,R2M,TPI,DA,DB,DF,DC,DIS,ZDIS,Y52
      COMMON/II/IP,MC,NS,JC,KC,MPRINT,NP,IDST,IINT
      WRITE(6,100)TAU,Y(4)
  100 FORMAT(2D15.7)
      RETURN
      END
```

```
      SUBROUTINE OUTP1
      IMPLICIT REAL*8(A-H,O-Y)
      DIMENSION ZT(500),ZY42(500),ZF2(500),ZR2(500),ZDIS(500)
      COMMON OMN,DZETA,OMN2,Y12,Y22,Y32,Y42,F2,R,T,TIME,R2,ZT,ZY42,ZF2,
     1ZR2,Y42M,R2M,TPI,DA,DB,DF,DC,DIS,ZDIS,Y52
      COMMON/II/IP,MC,NS,JC,KC,MPRINT,NP,IDST,IINT
      MC=MC+1
      DIFF=R2-Y42
      IF(IINT.EQ.2)DIFF=R2-Y52
      Y42P=Y42
      IF(IINT.EQ.2)Y42P=Y52
  100 FORMAT(6D11.4,I7)
  350 IF(MPRINT.EQ.1)WRITE(6,100)TIME,Y12,Y22,Y32,Y42P,F2,MC
      JC=JC+1
      IF(ICON.EQ.1)GO TO 450
      IF(JC.EQ.NS)GO TO 450
      RETURN
  450 KC=KC+1
      ICON=0
      IF(KC.GT.NP)RETURN
      ZT(KC)=TIME
      ZY42(KC)=Y42
      IF(IINT.EQ.2)ZY42(KC)=Y52
      IF(KC.EQ.1)ZF2(KC)=F2
      IF(KC.GT.2)ZF2(KC-1)=F2
      IF(KC.GT.2)ZF2(KC)=F2
      IF(IDST.EQ.0)GO TO 451
      DIS2=DA+DB*TIME+DC*DSIN(DF*TPI*TIME)
      ZDIS(KC)=DIS2
  451 ZR2(KC)=R2
      JC=0
      IP=0
      RETURN
      END
      SUBROUTINE OUTP2(TAU,X,DERX,IHLF,NDIM,PRMT)
      IMPLICIT REAL*8(A-H,O-Y)
      DIMENSION ZT(500),ZY42(500),ZF2(500),ZR2(500),ZDIS(500),X(2),
     1DERX(2),PRMT(5)
      COMMON OMN,DZETA,OMN2,Y12,Y22,Y32,Y42,F2,R,T,TIME,R2,ZT,ZY42,ZF2,
     1ZR2,Y42M,R2M,TPI,DA,DB,DF,DC,DIS,ZDIS,Y52
      COMMON/II/IP,MC,NS,JC,KC,MPRINT,NP,IDST,IINT
      MC=MC+1
      DIFF=R-X(2)
      IF(MPRINT.EQ.1)WRITE(6,100)TAU,Y12,Y22,X(1),X(2),F2,MC
  100 FORMAT(6D11.4,I7)
      JC=JC+1
      IF(ICON.EQ.1)GO TO 450
      IF(JC.EQ.NS)GO TO 450
      RETURN
  450 KC=KC+1
      ICON=0
      IF(KC.GT.NP)RETURN
      ZT(KC)=TAU
      ZY42(KC)=X(2)
      IF(KC.EQ.1)ZF2(KC)=F2
      IF(KC.GT.2)ZF2(KC-1)=F2
      IF(KC.GT.2)ZF2(KC)=F2
      IF(IDST.EQ.0)GO TO 451
      DIS=DA+DB*TAU+DC*DSIN(DF*TPI*TAU)
      ZDIS(KC)=DIS
  451 ZR2(KC)=R
      JC=0
      IP=0
      RETURN
      END
      SUBROUTINE ULINE(X,Y,I)
      IF(I.EQ.1)GO TO 100
      CALL DRAWA(X,YOLD)
      CALL DRAWA(X,Y)
  100 YOLD=Y
      RETURN
      END
```

Index